Instructor's Manual to Accompany

Design and Equipment for Restaurants and Foodservice

A Management View

Third Edition

Costas Katsigris and Chris Thomas

WILEY

JOHN WILEY & SONS, INC.

This book is printed on acid-free paper. ♾

Published by John Wiley & Sons, Inc., Hoboken, New Jersey.

Published simultaneously in Canada.

For general information on our other products and services, or technical support, please contact our Customer Care Department within the United States at 800-762-2974, outside the United States at 317-572-3993 or fax 317-572-4002.

Wiley also publishes its books in a variety of electronic formats. Some content that appears in print may not be available in electronic books. For more information about Wiley products, visit our Web site at http://www.wiley.com.

Library of Congress Cataloging-in-Publication Data:

ISBN: 978-0-470-25401-1

Printed in the United States of America.

10 9 8 7 6 5 4 3 2 1

TABLE OF CONTENTS

TO THE INSTRUCTOR:

We have assembled this Instructor's Manual to accompany the text *Design and Equipment for Restaurants and Foodservice, Third Edition.* It contains the following:

COURSE OUTLINES

Flexibility is as important in teaching about foodservice as it is in the industry itself. So, we offer several different course outlines, to meet the needs of different semester lengths and/or to emphasize different topics. These options include:

- A basic, full-credit, one academic semester course that covers 15-16 weeks and focuses on facilities management and major types of restaurant equipment. This course is based on using Chapters 1 through 16, in order.
- An alternative 15-16 week course that focuses on major types of restaurant equipment, plus smallware, tabletop components, and linens. This includes Chapters 1 through 4, and 9 through 20, and omits the discussion (in Chapters 5 through 8) about utilities, environmental issues, safety, and sanitation.
- A basic course for a 10-week academic quarter, which focuses on facilities management and major types of restaurant equipment – again, utilizing Chapters 1 through 16.
- An alternative 10-week course, which focuses on all major equipment categories, plus smallware, tabletop components, and linens. This path also omits the utility-related chapters (5 through 8), and requires that more than one chapter be covered in some weeks.

In each outline, we have highlighted chapter content and keyed the outline to the text with page references.

CHAPTER SUMMARIES

On the pages facing the outlines, we have listed resource suggestions pertinent to each lesson: Ideas for guest speakers, helpful trade publications, and so on. . Use your imagination and thoroughly research your local and regional resources and your students will be the beneficiaries.

After the outlines and "Outside Resources" sections, you'll find a short summary of the main points discussed in each textbook chapter. This format is intended to help you present an enormous amount of material in an organized fashion. Use our ideas as guidelines, and add your own. The areas of foodservice design and equipment are wide open to interpretation, and we hope your students will appreciate the ability to design and equip their own imaginary facilities in your class.

For each chapter, there are:

- **Teaching Goals** - Brief statements of the topics we think students should learn from the chapter. You will, of course, want to expand on these ideas.
- **Major Points of Emphasis** - An extension of the teaching goals, these summarize or elaborate on the main ideas found in the chapter.
- **Points to Ponder** - Discussion questions are found at the end of each chapter in the textbook. Here, we point you to the information and page numbers in the text to answer

each discussion question. Again, the questions can be excerpted for use as essay questions, or simply discussed in class.

- **Quiz Questions** - These are tied to the major points of emphasis, and organized so that they can be photocopied for use as tests. The addition of an Answer Key enables you to create a quick master, and machine-grade the finished quizzes in a matter of minutes. In addition to test use, the lists of questions also come in handy to spark end-of-chapter discussions and/or pre-test reviews.

LOOKING BEYOND THE CLASSROOM

We would suggest a highly interactive format, using as many actual examples from the hospitality industry as can be gathered. Even for the topics that aren't as glamorous as some other areas of your program, we've tried to provide suggestions.

As we have observed over the years, the foodservice industry has become more mature and more sophisticated, a trend that offers a richer pool of resources for academia. We cannot stress strongly enough the need to relate the textbook content to the "real world" of food, beverage, and hospitality management. You will find many professionals in these fields who are more than willing to visit your classroom and share their stories with students.

However, it is up to you to be considerate of their time. In our experience, a visiting speaker needs as much as four weeks' notice, and a follow up (by e-mail or letter) confirming the date, time, and place he or she is being asked to appear. In addition, the guest should understand the level of discussion expected, and should know something about the students who will be addressed. A copy of the course outline would be helpful. Ask well in advance if the speaker intends to bring handouts and visual aids, and be sure you can accommodate them with the necessary hardware: a CD player, computer for PowerPoint presentations, and so on.

A reminder – one week before the classroom appearance, and again the day before – is smart. Many of these people are extremely busy, so have an alternate plan, or be prepared to fill the class time yourself in case of last-minute cancellation. Be sure your guest's appearance is in compliance with your school's guest speaker policy.

Also, do not overlook the resources on your own school campus. How do the foodservice professionals in your school's dining facilities budget, plan for, and shop for equipment? Colleagues in departments such as interior design, architecture, and construction trades are excellent sources. Just be sure they are aware, well in advance, of the emphasis and level of complexity you require about their particular topic. If your school teaches such topics as Architecture or Interior Design, consider a joint project with students from both departments.

'ACTION PLAN' PROJECTS

One idea that has worked especially well for Katsigris' classes over the years is an all-encompassing class project we refer to as an "Action Plan." It is included near the end of this manual. Action Plans can be assigned in any of several ways:

- Individual students work independently on their own projects.
- Small groups of 3 to 5 students work on the project as a group.
- Hospitality Management students are paired with Interior Design students and both professors oversee the collaboration of students in the two departments.

READING LIST AND ONLINE RESOURCES

Finally, we've included a list of periodicals, journals, and Web sites. The Internet is a rich source of material, both for you and your students. You'll find that most foodservice equipment manufacturers have Web sites full of information. We urge you to take some time and search for pertinent sites, and to assign online research projects to your students.

Local, state, and national organizations, including the National Restaurant Association, the North American Food Equipment Manufacturers Association (NAFEM), your local health inspectors, and others, are also excellent outside resources for this course. More than 700 equipment manufacturers are NAFEM members, so its Web site alone (www.nafem.org) is a wealth of information.

It seems every time we turn around, yet another restaurant sprouts up, even as another closes. Yes, it's a tough business – but one that continues to blossom, even in times of economic downturn, for those who know their markets, know their limits, and stick to their budgets. Being able to teach individuals to succeed in this very challenging industry is an exciting experience, and we trust that our textbook will help you do a credible job.

— Costas "Gus" Katsigris and Chris Thomas

COURSE OUTLINES

BASIC COURSE OUTLINE - FULL ACADEMIC SEMESTER

(Sixteen 3-hour sessions, or 48 student contact hours)
PATH ONE - A general foodservice course

SESSION	SUBJECT	TEXT PGS
1	Dining-out trends and observations. Developing and fine-tuning a restaurant concept. Determining financial feasibility with market and location research. Buying versus leasing a site, and common restaurant lease terms.	**Ch. 1 1-28**
2	Restaurant atmosphere and design. Components of restaurant atmosphere, including comfort and safety. Working with design professionals. Designing multi-use areas and public spaces. Tables and chairs. Bar, banquet, and private dining areas. Mobile foodservice options.	**Ch. 2 29-61**
3	Principles of kitchen design. Budgeting for commercial kitchen construction. Goals of restaurant service system, including flow patterns. Kitchen space analysis. Designing a banquet kitchen. Food Establishment plan Review Guide and HACCP compliance considerations.	**Ch. 3 64-97**
4	Space allocation, front, and back of house. Planning public and outdoor areas, including parking lots. Dining areas, kitchen areas, and auxiliary space needs. ADA considerations.	**Ch. 4 98-123**
5	Energy use and management in foodservice. Basic electricity terminology and measurement. How electrical systems work. Reading an electric bill. Tips for energy consumption and conservation. Alternative energy sources and "green" buildings.	**Ch. 5 126-162**
6	The use of gas, steam, and water in restaurants. Gas energy, terminology, how meters work, and energy savings. Steam energy requirements, terminology, and common problems. Water quality factors and water use. Discharge systems; hot water for restaurants	**Ch. 6 163-203**
7	The importance of lighting and types of fixtures. Noise and sound control. Heating, air conditioning, and kitchen ventilation. Indoor air quality and technology advances.	**Ch. 7 206-240**
8	Fire protection systems. Ergonomics and worker comfort and safety. Flooring materials, cleaning, and maintenance. Cleaning stainless steel. Sanitation and safe food handling, including NSFI, and HACCP standards. Waste management: reduction, reuse, recycling, composting, incineration, and landfill use.	**Ch. 8 241-266**
9	Foodservice equipment purchases. Gas versus electric appliances. Determining total cost of ownership (TCO) and depreciation. Making used and/or online equipment purchases. Leasing equipment. Writing equipment specifications. Warranties and warranty fulfillment. Equipment installation, start-up, and safety. After-purchase maintenance and repair considerations.	**Ch. 9 270-305**
10	Equipment needs in receiving and dry storage areas: scales, carts, trucks, and shelving. Determining needs for refrigerated storage. How refrigeration systems work. Walk-in coolers, freezers, specialty refrigeration units, such as draft beer coolers and soft-serve machines. Determining ice needs and selecting/maintaining	**Ch. 10 306-333**

	an ice maker.	
11	Ranges and ovens. Basic principles of heat. Gas and electric range tops Types of ranges and ovens and their common uses. New oven technology.	**Ch. 11 337-367**
12	Fryers and fry stations. How fryers work, both gas and electric. Care and conservation of frying oil. Pressure fryers. Fry station components. Pasta cookers. Technological advances in frying.	**Ch. 12 368-389**
13	Broilers, griddles, and tilting braising pans. Types and uses of broilers. Types and uses of griddles. Maintenance tips. Installation and use of tilting braising pans.	**Ch. 13 392-413**
14	Steam cooking equipment. How steam works. Steam jacketed kettles. Pressure steamers and pressureless steamers. Combination steamer / ovens. Steam generators. Cleaning and maintenance of steam equipment.	**Ch. 14 414-437**
15	Cook-chill technology. Explanation of all steps in the cook-chill process, including blast chilling, tumble chilling, and other chilling options. Storage and distribution of cook-chilled foods. Rethermalization of foods.	**Ch. 15 438-454**
16	Miscellaneous kitchen equipment: mixers and attachments, slicers, choppers, and food processors. Blenders, toasters, and food warmers. Coffee makers and espresso machines.	**Ch. 17 485-507**

ALTERNATE COURSE OUTLINE - FULL ACADEMIC SEMESTER

(Sixteen 3-hour sessions, or 48 student contact hours)

PATH TWO - Emphasis on Equipment, Smallware, and Tabletop

SESSION	SUBJECT	TEXT PAGES
1	Dining-out trends and observations. Developing and fine-tuning a restaurant concept. Determining financial feasibility with market and location research. Buying versus leasing a site, and common restaurant lease terms.	**Ch. 1** **1-28**
2	Restaurant atmosphere and design. Components of restaurant atmosphere. Design of individual areas. Considerations in dining room, including chair and table selection and decorating trends. Bar area design. Banquet rooms.	**Ch. 2** **29-61**
3	Designing the kitchen. Overall placement of kitchen. Service system of the restaurant, including flow patterns. Kitchen space analysis. The banquet kitchen.	**Ch. 3** **64-97**
4	Space allocation, front, and back of house. Planning public areas and outdoor areas, including parking lots. Dining areas, kitchen areas, and auxiliary space needs. ADA considerations.	**Ch. 4** **98-123**
5	Purchasing foodservice equipment. Selecting essential equipment. Buying used equipment; online purchases. Leasing equipment. Writing equipment specifications. Equipment installation and safety rules.	**Ch. 9** **270-305**
6	Receiving area equipment needs. Determining refrigeration needs. Walk-in coolers, freezers, specialty refrigeration units. Soft-serve machines and ice makers.	**Ch. 10** **306-333**
7	Ranges and ovens. Basic principles of heat. Gas and electric range tops. Types of ranges and ovens and their common uses. New oven technology.	**Ch. 11** **337-367**
8	Fryers and fry stations. How fryers work. Gas and electric fryers. Care and conservation of frying oil. Pressure fryers. Fry station components. Pasta cookers. Technological advances in frying.	**Ch. 12** **368-389**
9	Broilers, griddles, and tilting braising pans types and uses of broilers. Types and uses of griddles. Maintenance tips. Installation and use of tilting braising pans.	**Ch. 13** **392-413**
10	Steam cooking equipment. How steam works. Steam jacketed kettles. Pressure steamers and pressureless steamers. Combination steamer/ovens. Steam generators. Cleaning and maintenance of steam equipment.	**Ch. 14** **414-437**
11	Cook-chill technology. Explanation of all steps in the cook-chill process, including blast chilling, tumble chilling, and other chilling options. Storage and distribution of cook- chilled foods. Rethermalization of foods.	**Ch. 15** **438-454**
12	Dishwashing and waste disposal. Types and sizes of dishwashing systems. The use of booster heaters. Food waste disposers, waste pulpers, and collectors. Pot and pan washing.	**Ch. 16** **457-484**
13	Miscellaneous kitchen equipment: mixers and attachments, slicers, choppers, and food processors. Blenders, toasters, and food warmers. Coffee makers and espresso machines.	**Ch. 17** **485-507**

14	Kitchen smallware, tools, and equipment. Hand tools, kitchen cutlery, measuring tools. Pots, pans, and specialty bakeware. Serving and holding containers.	**Ch. 18** **508-533**
15	Tabletop items: Purchasing, cleaning and care of plateware, flatware, glassware. Common manufacturing processes for these items.	**Ch. 19** **536-557**
16	Linens and table coverings. How fabrics are woven and used; textiles and fabric alternatives. Care and cleaning of napery. On-premise laundry equipment and troubleshooting laundry-related problems.	**Ch. 20** **558-578**

BASIC COURSE OUTLINE – ONE ACADEMIC QUARTER

(Ten 3-hour sessions, or 30 student contact hours)

PATH ONE - Emphasis on Facilities and Equipment Management.

SESSION	SUBJECT	TEXT PGS
1	The economics of site selection. Concept development. Market feasibility and financial feasibility studies. Site selection. Acquisition of space; buying or leasing your site.	**Ch. 1** **1-28**
2	Restaurant atmosphere and design. Components of restaurant atmosphere. Design of individual areas. Considerations in dining room, including chair and table selection and decorating trends. Bar area design. Banquet rooms.	**Ch. 2** **29-61**
3	Designing the kitchen. Overall placement of kitchen. Service system of the restaurant, including flow patterns. Kitchen space analysis. The banquet kitchen.	**Ch. 3** **64-97**
4	Space allocation, front, and back of house. Planning public areas and outdoor areas, including parking lots. Dining areas, kitchen areas, and auxiliary space needs. ADA considerations.	**Ch. 4** **98-123**
5	Energy management. How energy is used in foodservice. Energy Audits. Measuring electrical energy. How electrical systems work. Energy consumption and conservation. Constructing energy-efficient buildings. Future energy outlook. The use of gas, steam, and water in restaurants. Terminology and meter reading. Water supply and quality. Plumbing fixtures and discharge system.	**Ch. 5-6** **126-203**
6	Environmental concerns, safety and sanitation. Ergonomic considerations. Lighting. Noise and sound control. HVAC and fire protection. Indoor air quality. Employee comfort and safety. Food safety and waste management issues.	**Ch. 7-8** **206-266**
7	Purchasing foodservice equipment. Making used and/or online purchases. Leasing equipment. Writing equipment specifications. Equipment installation and safety rules. Storage equipment, receiving area, and refrigeration needs. Walk-in coolers and freezers. Soft-serve machines, beverage dispensing units, and ice makers.	**Ch. 9-10** **270-333**
8	Ranges and ovens. Basic principles of heat. Gas and electric range tops. New appliance technology. Fryers and fry stations. Care and conservation of frying oil. Broilers, griddles, and tilting braising pans; installation, use and maintenance.	**Ch. 11-13** **337-413**
9	Steam cooking equipment. How steam works. Steam jacketed kettles. Pressure steamers and pressureless steamers. Combination steamer/ovens. Steam generators. Cleaning and maintenance of steam equipment. Cook-chill technology. Blast and tumble chilling. Safe storage and rethermalization of cook-chilled foods.	**Ch. 14-15** **414-454**
10	Dishwashing and waste disposal. Types and sizes of dishwashing systems. The use of booster heaters. Food waste disposers, waste pulpers, and collectors. Pot and pan washing. Miscellaneous equipment, including mixers and attachments, food processors and slicers, toasters, warmers, coffee makers, and more.	**Ch. 16-17** **457-507**

ALTERNATE COURSE OUTLINE – ONE ACADEMIC QUARTER

(Ten 3-hour sessions, or 30 student contact hours)

PATH TWO - Emphasis on Equipment, Smallware, and Tabletop.

SESSION	SUBJECT	TEXT PGS
1	The economics of site selection. Concept development. Market feasibility and financial feasibility studies. Site selection. Acquisition of space; buying or leasing your site.	**Ch. 1** **1-28**
2	Restaurant atmosphere and design. Components of restaurant atmosphere. Design of individual areas. Considerations in dining room, including chair and table selection and decorating trends. Bar area design. Banquet rooms.	**Ch. 2** **29-61**
3	Designing the kitchen. Overall placement of kitchen. Service system of the restaurant, including flow patterns. Kitchen space analysis. The banquet kitchen.	**Ch. 3** **64-97**
4	Space allocation, front, and back of house. Planning public areas and outdoor areas, including parking lots. Dining areas, kitchen areas, and auxiliary space needs. ADA considerations.	**Ch. 4** **98-123**
5	Purchasing foodservice equipment. Selecting essential equipment. Buying used equipment. Leasing equipment. Writing equipment specifications. Equipment installation and safety rules. Receiving area equipment needs. Food storage and shelving. Refrigeration equipment. Soft-serve machines and icemakers.	**Ch. 9-10** **270-333**
6	Ranges and ovens. Basic principles of heat. Gas and electric range tops. Types of ranges and ovens and their common uses. New oven technology. Fryers and fry stations. Care and conservation of frying oil. Broilers, griddles, and tilting braising pans; installation, use, and maintenance.	**Ch. 11-13** **337-413**
7	Steam cooking equipment. How steam works. Steam jacketed kettles. Pressure steamers and pressureless steamers. Combination steamer/ovens. Steam generators. Cleaning and maintenance of steam equipment. Cook-chill technology. Blast chilling and tumble chilling. How to store, distribute, and reheat cook-chilled foods.	**Ch. 14-15** **414-454**
8	Dishwashing and waste disposal. Types and sizes of dishwashing systems. The use of booster heaters. Food waste disposers, waste pulpers, and collectors. Pot and pan washing.	**Ch. 16** **457-484**
9	Miscellaneous kitchen equipment: mixers and attachments, slicers, choppers, and food processors. Blenders, toasters, and food warmers. Coffee makers and espresso machines. Smallwares, including knives, hand tools, and measuring tools, pots and pans, serving, and holding containers.	**Ch. 17-18** **485-533**
10	Tabletop items: Purchasing, cleaning, and care of plateware, flatware, and glassware. Common manufacturing processes for these items. Linens and table coverings, their manufacture, use and care. Determining napery needs for dining rooms, and suggestions for alternatives to fabrics. On-premise laundry equipment and maintenance.	**Ch. 19-20** **536-578**

OUTSIDE RESOURCES

Chapter	Suggestions for Guest Speakers; Sources of Ancillary Material
1	Your area Chamber of Commerce and Small Business Administration. Commercial real estate agents. Local planning and zoning officials. Marketing firms that specialize in restaurant development. For trend information, industry publications and the *New York Times.*
2	Architects and interior designers who specialize in restaurants. Foodservice consultants. Manufacturers' Web sites and catalogs for chairs and tables. Industry publications that focus on design and redesign of dining areas. An interior design instructor, possibly from your school. The magazine *Foodservice Equipment and Supplies*, which features monthly foodservice design layouts.
3	Foodservice consultants; equipment dealers and their design staff. The chef or owner of a new (or newly remodeled) restaurant, who can discuss service systems, show blueprints, and an equipment schedule. Industry publications that feature design and redesign plans of restaurant kitchens.
4	An architect or consultant who designs restaurants. City officials who review and approve restaurant plans. A representative of a restaurant chain who is responsible for building and opening new units. An individual in the process of opening a first restaurant. Industry publications that feature outstanding examples of new facilities.
5	A utility company representative who is familiar with foodservice clients or who performs energy audits. An electrical contractor who does restaurant wiring and equipment installation. U.S. Department of Energy representative, or consultant who specializes in energy conservation and/or future energy sources.
6	Water utility representative. Plumbing contractor. Manufacturer's representative for hot water heaters or plumbing fixtures. Gas company representative who handles foodservice accounts. Distributor of steam-operated equipment. Manufacturers' literature about features and prices of water heaters and plumbing fixtures.
7	Lighting design expert. Local utility company representative. Ventilation expert who builds and installs kitchen exhaust systems; find a local member of ASHRAE. City sanitation department representative who approves restaurant plans. Recycling company or federal E.P.A. representative. Field trip to examine restaurant exhaust hoods.
8	Foodservice equipment consultant. Local restaurant equipment dealers and/or manufacturers' representatives. Independent foodservice equipment installers. A fire department representative to discuss kitchen fires and fire prevention.
9	Manufacturers' representatives for commercial refrigeration, shelving, and storage equipment. Catalogs of shelving, carts, refrigerators, and icemakers. An equipment dealer, to discuss installation, training, start-up, and maintenance issues. A visit to a restaurant equipment supply business would help illustrate quality factors: construction details, equipment sizes, options, and accessories.
10	Commercial walk-in installation company (it's a specialty, but most cities of any

size have them) to discuss sizing, insulation, maintenance, and service for walk-ins. Beverage distributors, for discussion on soft drink and beer dispensing systems. A field trip could focus on foodservice receiving areas, "quality control," and intake procedures for deliveries.

11 A visit to a commercial kitchen is highly recommended at this stage of the course. Ask chefs to describe what they like (or don't like) about their hot lines, focusing on oven and range top equipment. Large-scale food distributors (such as Sysco) employ chefs; equipment dealers (like Bargreen Ellingson) have test kitchens at some locations.

12 Equipment dealers or manufacturers' representatives of fryers. Equipment repair technicians. Manufacturers' catalogs and training CDs. A knowledgeable salesperson for frying oil or filtration systems to discuss oil use, maintenance, and oil deterioration factors. Field trip to a QSR to see fry stations hard at work.

13 Manufacturers' representatives or equipment dealers who sell broilers, griddles, and/or tilting braising pans. A visit to a foodservice equipment dealer's showroom is suggested at this point in the course.

14 Most manufacturers of steam equipment have an extensive selection of sales and training CDs that include details of construction, use, and maintenance, plus catalogs and spec sheets. At this point in the course, a visit to a hotel, school, or hospital that uses steam equipment is suggested. Your own school cafeteria may have steam equipment and a cafeteria manager willing to demonstrate it for your class. For boilers and generators, a commercial plumber and/or steam equipment repair technician (or manufacturer).

15 Ask equipment dealers to share names of cook-chill users; invite them to class and/or plan a field trip to see their system at work.

16 Manufacturers' reps for dishwashing systems or detergent manufacturers. The person in your region's largest hotel who supervises dishwashing operations. A local health department representative. Manufacturers' info about construction and care of dishwashing machines, and booster heaters, waste disposers, and waste pulpers. If your campus has a cafeteria, analyze its dishroom as an example. Arrange for students to spend an afternoon or evening washing dishes there, or at a local restaurant.

17 For the most part, these appliances are portable and could be brought to class by a restaurant owner or small-appliance repair technician. Ask a coffee bar owner to come to class for a visit, or have students meet there for a field trip. If students have work experience at a coffee bar, ask them to lead the class discussion.

18 A chef or culinary instructor in your own program would be a good guest speaker. Ask this person to bring a variety of smallware items to class, and discuss how to choose them, and how to work with and care for them. A restaurant supply store field trip is another good option.

19 Tabletop sales reps or interior designers in your area with foodservice experience; invite them to class with samples. Larger cities will have showrooms to visit; alternately, ask students to visit restaurants and report specifically on their tabletop choices, prices.

20 A fabric store owner or employee can come to class with samples. Manufacturers can supply colorful catalogs and brochures, but it's important that students be able to *feel* the fabrics. Students can practice napkin-folding ideas and techniques – have a contest for "best new fold." Tabletop designers or decorators with foodservice clients. A commercial laundry operator or laundry manager for a large restaurant or hotel. Linen rental service owner or manager.

GETTING ORGANIZED:
Chapter-by-Chapter Materials and Quizzes

CHAPTER 1: Economics of Site Selection

TEACHING GOALS

- Introduce the ideas of concept development: How to choose a concept, and the variables involved in creating a concept.
- Emphasize the role of the feasibility study, and learning about your competition. Cover in detail both market and financial feasibility issues.
- Cover in detail the guidelines for site selection. Stress each point as it impacts a specific restaurant location.
- Describe the alternatives: Owning a space, or leasing it. Make students aware of the specifics of leasing.

MAJOR POINTS OF EMPHASIS

Concept development: In creating a concept, discuss the role of image, and the importance of a written statement that describes the concept. Mention the influence of theme, name, and logo in developing a restaurant concept. Ask students to study their own favorite restaurants and report on concepts, logos, themes, and reasons for popularity.

Industry trends: It's important that students understand that foodservice is a "trendy" business, affected by everything from employment statistics to social fads. The book (on page 2) cites a 2004 Zagat® survey – ask students to add their own ideas about other social phenomena they believe will play into the industry in the near future.

Feasibility studies: The market feasibility study is a close look at potential customers, trade area, location, and direct competition. In financial feasibility studies, the so-called hard numbers (like potential costs versus income) must be explained and understood. Pages 10-12 and 14-17 offer guidelines.

Site selection: Discussion of the importance of location is critical here. Judge restaurants in your area by guidelines covered on pages 17-22.

Owning versus leasing space: Buying land and building on it is based on the economics of the individual project. Examine both tax considerations and bank financing options. Stress the importance of finding out cost per square foot, either for construction or in a lease. When leasing, look at specifics like base rates, percentage factors, and exactly what is (and is not) included in the lease: maintenance and repairs, insurance, who pays real estate taxes, and so on. Lease particulars are mentioned on pages 22-27.

POINTS TO PONDER
(Reference pages for discussion questions found at end of chapter.)

1. There is a brief discussion in the section about branding and concepts (page 5) that mentions "emotion," but students are on their own in formulating this answer, and should be encouraged to discuss restaurants they "feel good" about – and exactly why they feel that way.

2. Again, answers will depend on your local restaurant market. The four guidelines are listed on page 3.

3. Students are asked to identify local restaurants that are in the entrepreneurial, emerging, expanding, and mature stages of their businesses. Friedman's discussion centers on MCOs, with numbers of locations to gauge a chain's success at each level. (You might tell students they don't have to stick strictly to the "numbers" guidelines in making their selections.)

4. On average, $6,000 per day. (200 x 3 = 600; 600 x $10 = $6,000)

5. The purpose of market studies is explained on pages 10-12; for financial feasibility studies, on pages 14-17.

6. On page 21, the term *restaurant cluster* is defined. Students must look in their own community for a specific answer.

7. Pages 22-23 mention "two long-term factors" in a building a restaurant, the floating interest rate and payout period; but we're asking for three, so the list should include an item about being able to work successfully together with the design team.

8. Pages 18-21 list 16 different factors in restaurant site selection. Students are asked to choose four of them, and to make their own good/bad judgments about these.

9. As an instructor, you can learn a lot about your students by their answers to this question. There's no right or wrong, just individual preferences. Grade based on their ability to explain and justify their decision.

10. Pages 24-25 define these terms and offer examples.

QUIZ – CHAPTER 1

Circle the letter before each phrase that completes the statement correctly. If you are using a Scantron form, follow the directions on the form. Be aware that there is only one correct answer per question.

1. This socioeconomic change in the United States has affected the foodservice industry.
   (a) The school lunch program.
   (b) The rise of QSRs.
   (c) The use of the microwave oven.
   (d) The home meal replacement trend.

2. The term fast casual is:
   (a) Another way of saying QSR.
   (b) Applied to fast food concepts.
   (c) Can be applied to upscale restaurant concepts.
   (d) None of the above.

3. Concept development is a collection of variables that, together, make up what guests will see as the restaurant's__________.
   (a) Menu.
   (b) Logo.
   (c) Image.
   (d) Décor.

4. A restaurant's theme is part of its ______________.
   (a) Décor.
   (b) Entertainment.
   (c) Concept.
   (d) Design.

5. A crucial step in developing a concept is:
   (a) Kitchen layout.
   (b) Menu development.
   (c) Décor.
   (d) Marketing evaluation.

6. Branding, as it applies to restaurant concepts, involves ________________.
   (a) The logo.
   (b) The menu.
   (c) The creation of signature dishes.
   (d) All of the above.

7. An advantage of a non-traditional restaurant site is:
   (a) You can rely on the host site's business projections.
   (b) You may be judged by the company you keep.
   (c) It is faster to open this type of business.
   (d) The host site offers a built-in method of quality control.

8. The written document prepared to determine the potential success of a restaurant concept is called its _____________.

(a) Feasibility study.
(b) Mission statement.
(c) Demographic study.
(d) Psychographic study.

9. An example of a fast casual dining concept is the _____________.

(a) Doughnut shop.
(b) Café/bakery.
(c) Gourmet frozen yogurt shop.
(d) All of the above.

10. As we consider the cyclical nature of a restaurant business, at what point is its brand name the strongest?

(a) In its mature stage.
(b) In its entrepreneurial stage.
(c) In its emerging stage.
(d) In its expansion stage.

If a statement below is entirely true, circle the letter 'T' in front of it. If it is NOT ENTIRELY true, circle the letter 'F.' If you are using a Scantron form, follow the directions on the form.

11. T F The primary reason for opening multiple locations of a restaurant is to protect the existing business.

12. T F For restaurants, *convenience concepts* meet the needs of consumers who, for various reasons, cannot eat at the restaurant.

13 T F The *life cycle* of your customer base remains constant.

14. T F *Psychographics* is a term that refers to the personal habits of a restaurant's potential customers.

15. T F The crime rate in the area surrounding a prospective restaurant site is part of the information found in a *market feasibility study.*

16. T F Most lenders expect a prospective restaurateur to invest at least 20 percent of the capital costs for start-up from personal sources.

17. T F A *limited liability corporation* is created in order to allow a large group of investors to fund a restaurant concept.

18. T F *Prime costs* include the costs of rent, labor, food, and beverages.

19. T F A *triple net lease* includes rent, real estate taxes, and some forms of building insurance in a single, monthly payment.

20. T F In a restaurant lease, an *escalation clause* is an automatic increase in the amount of the monthly rent after an agreed time period, usually five years.

CHAPTER 2: Restaurant Atmosphere and Design

TEACHING GOALS

- For this chapter, as well as Chapters 3 and 4, we suggest each student select a particular facility. It may be his or her dream restaurant of the future, or an existing eatery in your community. That way, the student can apply the principles in a practical sense, to a specific location.
- Discuss the key components of atmosphere in a foodservice setting. Talk about ways to five customers a welcome change of pace. Discuss ways that guests' comfort and security play a role in determining atmosphere.
- Discuss redesign of existing space, and when an owner should consider a redesign.
- Examine the step-by-step process a guest goes through from the time he or she first notices the restaurant's signage, to waiting for a table and being seated.
- Look at mobile foodservice options like kiosks and food carts.
- Discuss the pros and cons of having a bar area or cocktail lounge.
- Discuss what to consider (comfort, function, design, size, adaptability) when choosing tables and chairs.
- Stress the versatility of a well-designed banquet room.

MAJOR POINTS OF EMPHASIS

Ambiance and atmosphere: What is necessary to create them? Introduce the types of professionals who might help in creating them: architects, interior designers, and foodservice consultants. Invite these folks to be part of your classroom discussions, and ask them to bring visuals. Also discuss the pros and cons of doing it yourself, without hiring these professionals. Trade publications often feature the latest design trends.

Elements of atmosphere: Discuss how to incorporate comfort and security into design. What makes people feel safe? What makes them feel comfortable? Use the guidelines on page 29-30, and page 32. Telltale signs that redesign is necessary for an existing business are listed on pages 35-36.

Designs of specific spaces and their overall importance: Building exteriors; the entryway and its influence on guests. Dining room layout and flow patterns. Location and design of restrooms. Laying out bar areas and the use of portable bars. Students can use the *Follow the Guest* details on pages 41-44 and in other restaurants they visit.

Non-traditional alternatives: Mobile foodservice options are discussed on pages 44-47. Make sure students understand the demands of remote locations.

Bars and lounges: Including a bar area is covered on pages 47-52.

Furniture: Students can look at manufacturers' catalogs to emphasize points made about good quality construction and selection of tables and chairs, as well as the many different styles and fabrics available. Stress the versatility of smaller table sizes that can be pushed together as needed.

Banquet facilities: The banquet room (discussed on pages 55-59) should be among the most adaptable areas of a restaurant. Table shapes and sizes will help determine how many ways the room can be set up for different purposes. Sturdy construction details of furnishings are mentioned, as well as unique banquet accessories, such as platforms, partitions, and portable dance floors.

POINTS TO PONDER

(Reference pages for discussion questions found at the end of each chapter)

1. At the top of page 34, a text box that contains seven different components of ambiance. We ask that the student list three of them.

2. "Advice from Top Designers" is on page 60 of the book. This question could prompt a short report by each student on the philosophy of a different designer.

3. Pages 41-42 explain multiple ways in which entry space can be put to use.

4. Privacy, comfort, and attractiveness are all important, but the single most critical factor on which a restroom will be judged is its cleanliness. Details are on pages 43-44.

5. The idea here is that you should decide what (and how big a) role the bar will play in your business. This includes potential profit, how much space you have to work with, whether the bar will be a destination apart from other dining, and if you'll provide entertainment. Pages 49-51 mention the options.

6. The flow pattern of a dining area is the process of delivering food and beverages to customers.

7. The answers here could be amusing! What we're looking for is an honest assessment of what mature diners may be looking for, from better lighting, to fewer stairs, to more convenient parking.

8. Pages 53-54 offer tips about style, construction, and comfort of dining chairs.

9. Truthfully, you can only seat 84 people comfortably with this number and size of tables, not 86. Page 55 offers a list of tips, but the students are on their own to sketch their dining rooms. That's why we've asked them to describe their layouts.

QUIZ – CHAPTER 2

Circle the letter before each phrase that completes the statement correctly. If you are using a Scantron form, follow the directions on the form. Be aware that there is only one correct answer per question.

1. One good rule for developing a restaurant's atmosphere is to:
   (a) Use a pleasing mixture of colors.
   (b) Provide a change of pace for prospective guests.
   (c) Specify the sizes of every table.
   (d) Specify the types of seating at every table.

2. As it pertains to restaurants, environmental psychology involves:
   (a) Safety and security.
   (b) Comfortable lighting.
   (c) The use of large windows for natural light.
   (d) High ceilings to make rooms appear more spacious.

3. Which of the following characteristics should figure most prominently in selecting a design professional?
   (a) Size and age of the design firm.
   (b) Chemistry between owner, investor(s), and design team.
   (c) Project experience.
   (d) Fee structure.

4. Which of these can be said about the design program for a restaurant?
   (a) It details all the criteria for the restaurant's design.
   (b) It is a legal contract.
   (c) Its contents are the sole responsibility of the architect.
   (d) All of the above.

5. In planning a restaurant's atmosphere, this must be considered:
   (a) The changing demographics of the prospective guests.
   (b) Its ability to meet the needs of the various types of customers.
   (c) The need to create comfort for all guests.
   (d) All of the above.

6. Which one of the following statements *will not* solve a redesign dilemma?
   (a) Repainting the dining space in a different hue.
   (b) Installing window coverings where there have been none.
   (c) Installing solar panels for hot water heating.
   (d) Reupholstering the dining chairs.

7. This can be said about having a private dining room in a restaurant:
   (a) It may help control food costs.
   (b) It gives some control of the space to the host/guest who rents it.
   (c) It provides for privacy for some types of events and gatherings.
   (d) All of the above.

8. The flow pattern of a dining room:
   (a) Should take both guest and employee safety into account.
   (b) Should include the display of both foods and wines.
   (c) Is based on analysis of noise and light levels in the dining space(s).
   (d) Should take wall and floor coverings into account.

9. In foodservice facilities, restrooms:
   (a) Are usually located at the back of the dining area
   (b) Are usually based on location of water lines
   (c) Do not usually require temperature control
   (d) Are most often lit with fluorescent lighting

10. The difference between a cart and a kiosk is:
   (a) Kiosks are mobile.
   (b) Carts are mobile.
   (c) Carts are fixed.
   (d) Kiosks are only found outdoors.

If a statement below is entirely true, circle the letter 'T' in front of it. If it is NOT ENTIRELY true, circle the letter 'F.' If you are using a Scantron form, follow the directions on the form.

11. T F Food carts can be up to 20 feet in length.

12. T F The concession trailer typically used for foodservice (at venues such as ballparks and county fairs) has its own brakes and signal lights so it can be safely driven and/or towed.

13. T F Like any other commercial kitchen, mobile foodservice facilities require a separate hand-washing sink and three-compartment pot-washing sink.

14. T F For restaurant seating, the depth of the seat from the front edge of the chair to its back should be 16 inches.

15. T F The *brass rail* is part of the *backbar.*

16. T F The *front bar* includes the *pouring station(s).*

17. T F Square tables take up more space in a room than round tables.

18. T F A 36-by-36-inch square table will seat four guests comfortably.

19. T F A *glide* is a self-leveling device associated with banquet tables.

20. T F The most common folding tables used in banquet facilities measure 72 inches or 96 inches long, and 15 to 18 inches wide.

CHAPTER 3: PRINCIPLES OF KITCHEN DESIGN

TEACHING GOALS

- Emphasize the fact that the design budget must fit the concept, and that kitchen space is the most expensive part of the design. Include timelines as part of the budget and design process.
- Discuss in detail the service system and its influence on product and traffic flow. Illustrations are provided for various flow plans used in kitchens.
- Experiment in class with the use of bubble diagrams to plan space and point out relationships of different work sections and work centers of the kitchen. Look at specific equipment needs for work centers.
- Stress the critical role of the service area (wait station) and how it differs in different types of foodservice businesses.
- Break the food preparation area and production areas down into their basic components, and discuss each function within the area.
- Discuss practical considerations about warewashing, kitchen doors, work surfaces, flooring, and so on. In some instances, it really is "the little things" that make a difference in how efficiently and safely a restaurant kitchen can operate.
- Mention a restaurant's banquet options, including ways to make the space as versatile as possible.

MAJOR POINTS OF EMPHASIS

Teach students to match design and concept to budget and space. Cost outlays should be balanced between front and back of the house. (Use the worksheet on page 60 of the book.) Timelines are just as important as staying on budget.

First decisions to be made: Where to place that kitchen and what service system will be used to deliver food to the guests. From there, flow patterns emerge, and other areas can be placed based on necessary proximity to the kitchen. Use the bubble diagram to map out workstations.

Flow patterns are just as important within the kitchen as in the dining area. Pages 74-80 describe and illustrate some common flow patterns.

If a display kitchen is decided on, discuss the additional planning, design, and staff training details that will be required.

Discuss the importance of the service area or wait station. How many should there be, and what functions should they include? How does the use of a pass window and an expediter work in tandem with wait staff, and the other wait stations?

A *Human Engineering Checklist* is included on page 73, which can be used to begin discussions (to continue in other chapters) about comfort and safety for employees, not just for guests.

Cafeterias require unique flow patterns (serving lines), which are shown on pages 81-82. Explain the food preparation and production areas, and the work done in each; these are described on pages 82-88. Pay special attention to the production areas, including cooking duties, pantry, and bakery.

Stress the health impact or proper warewashing, and the importance of design in this area to promote sanitation and minimize the breakage.

Discuss the small details that make things run more smoothly in a kitchen: Everything from choices of easy-to-clean materials for floors and walls, to proper worktables and landing spaces for heavy trays of food, to placement of trash cans.

And finally, discuss the service kitchen in a banquet area as an extension of the main kitchen, with similar but separate requirements. Talk about where to locate it, its size, and what types of equipment are needed there.

POINTS TO PONDER

(Reference pages for discussion questions found at the end of each chapter)

1. There are enough items listed on pages 71-72 to prompt students to create a short report on how kitchen location affects each of them. We stress this aspect of design because so many other considerations are determined by where the kitchen is placed in the building.

2. Students may point out that this also was asked at the end of Ch. 2, and they are correct. However, here, this question deals with the two types of flow patterns in a commercial kitchen (explained on pages 74-75). Another flow pattern involves pick-up and delivery of food and beverages to guests.

3. Definitions for these three areas (fabrication, preparation, and production) are found on pages 82-83.

4. Possibly the least attractive, but some of the most necessary, the pot sink, the wait station, and the warehousing area are among the most often ignored. (Students will really have to scour the chapter to find these answers, but they're there!)

5. Page 93 lists the basic service kitchen equipment. Students can make their own judgments about what might be shared (on carts with rolling casters?) with the main kitchen.

6. Yes, the counters and beverage areas where customers get food and drinks.

7. On the hot line, space is allocated based on the types of cooking you will do (high-volume cooking in batches, or individual orders), and the cooking methods (which require different types of equipment). In the pantry, the amount of refrigeration you'll need and the types of cold food you prepare will determine space allocation. This is explained on pages 84-89.

8. Ventilating hoods and an automatic fire extinguishing system. The reasons are explained on page 84.

9. They are placed in special racks or containers for the wash process, and they shouldn't be towel-dried after washing.

10. Page 91 mentions considerations for door safety; things like swinging double doors, clearly labeled IN and OUT so that people don't run into each other when things get busy. These apply to areas other than dishrooms, as well.

QUIZ – CHAPTER 3

Circle the letter before each phrase that completes the statement correctly. If you are using a Scantron form, follow the directions on the form. Be aware that there is only one correct answer per question.

1. The detailed arrangement of the kitchen floor and counter space is known as its:
   (a) Design.
   (b) Bulletin.
   (c) Layout.
   (d) Work centers.

2. Which of the following is design rather than layout?
   (a) The relationship among functional areas.
   (b The configuration of equipment in a kitchen.
   (c) The arrangement of a cooking island.
   (d) The locations of hand sinks throughout a kitchen.

3. An example of a work section is a:
   (a) Potsink.
   (b) Broiler.
   (c) Cold food area or pantry.
   (d) Exhaust hood.

4. This can be said about a cooking suite:
   (a) It is freestanding.
   (b) It is designed for specific pieces of equipment.
   (c) It is most popular in QSR locations.
   (d) It is located against a wall of the kitchen.

5. In budgeting for a new kitchen, the recommended contingency amount to be included in the cost is:
   (a) 10 percent.
   (b) 20 percent.
   (c) 30 percent.
   (d) None of the above.

6. When planning a new kitchen, the term rough-ins refers to:
   (a) Utility connections.
   (b) Locations of exhaust hoods.
   (c) Estimates of aisle dimensions.
   (d) Heights of the pieces of equipment.

7. Space requirements for a specific kitchen work station are influenced by:
   (a) The number of persons working in this part of the kitchen.
   (b) The amounts and types of equipment required in a specific area.
   (c) The types of products to be produced in this work area.
   (d) All of the above.

8. A restaurant's service system is:
   (a) The types of plates and flatware used to serve food.
   (b) How tables and booths are arranged in the dining area.
   (c) How food and beverages are delivered to the guest.
   (d) Whether a point-of-sale system is used.

9. The pass window is found in:
   (a) The delivery area.
   (b) The kitchen service area.
   (c) The wait station.
   (d) The dining room.

10. The distance from kitchen to dining room determines which of these?
   (a) The number of wait stations.
   (b) The activities that must take place in the wait stations.
   (c) Flow patterns to and from kitchen and dining areas.
   (d) All of the above.

If a statement below is entirely true, circle the letter 'T' in front of it. If it is NOT ENTIRELY true, circle the letter 'F.' If you are using a Scantron form, follow the directions on the form.

11. T F In fine dining establishments, there is little food handling that takes place at the wait stations.

12. T F The *ingredient room* is part of the storage area of a commercial kitchen.

13. T F When materials are passed from one *work station* to another, it is most efficient to use a *straight line* arrangement.

14. T F When space is not sufficient for a *straight line* or *parallel* arrangement, an *L-shaped* kitchen design is used.

15. T F The dishroom requires special attention from engineers to control humidity.

16. T F An *expediter* facilitates food ordering and the delivery of cooked food in a busy commercial kitchen.

17. T F If double doors are used between a kitchen and dining area, they must be 30 inches wide.

18. T F In cafeterias, the *shopping center* arrangement makes it easy for guests to bypass any section.

19. T F The *garde manger* is an area of the commercial kitchen where cold foods are prepared.

20. T F In the *forming area* of a *baking section*, items made of dough (rolls, pies, and so on) are placed in pans.

CHAPTER 4: SPACE ALLOCATION

TEACHING GOALS

- Discuss restaurant space planning as a sequence of events. Stress the fact that there are important decisions to be made prior to the design process, like what type of food will be served and how it will be served; and the necessity of planning on paper (or by computer).
- The preliminary space planning stage includes all public areas, from parking and outdoor dining, to the entryway, dining area, banquet room, service areas, bar, and restrooms. This chapter mentions some of the unique considerations for each of these areas.
- Building on Chapter 2's background about table and chair options. More information is given about arranging these in the dining area. Use the sample floor plans included in the text.
- Review the kitchen design points made in Chapter 3 as you acquaint students with suggestions for allocating kitchen space. Ask students to debate the necessity of the various auxiliary areas (receiving, office, locker rooms, etc.) where space is tight.
- Outline process for making a facility compliant with the Americans With Disabilities Act. (In mid-2008, the ADA was updated by Congress – check online resources for information relating to these updates.)

MAJOR POINTS OF EMPHASIS

There are different *space allocation considerations for exterior features* (paring, patio dining); front of house (dining areas, bars, wait stations, entryways); back of house (kitchen, dishwashing); and auxiliary areas (receiving, storage, offices, etc.).

Preliminary space planning decisions include market analysis of potential customers, type of food and service system. First, sketch out on paper the relationships among spaces; then, make a preliminary layout sketch. Include not just space names, but locations of equipment.

In outdoor spaces, city ordinances may regulate parking safety issues. There may be ordinances about acceptable sizes parking spaces, entrances and exits, and how much parking must be provided depending on the size of the building or number of guests it can hold. For outdoor dining, there are special considerations like how to provide water or lighting in these spaces, or where to place wait stations.

Front of the house considerations build on your preliminary space plans. You will require very different amounts of space per guest depending on the type of dining. Pages 98-101 offer several basic guidelines, along with sample floor plans. In giving students space planning assignments, be sure to ask them to use a variety of table sizes, booths, and banquettes. For bar stools, stress the importance of bar placement and stool selection.

Restrooms are another area of multiple legal requirements, from numbers of toilet facilities for men and women. And of course, they must be accessible to people with disabilities.

Review the Chapter 3 discussion of kitchen area necessities. Take the dishroom layout one more step, with the placement of dish tables and the dishwashing machine.

Receiving areas should be planned and located based on volume of goods received and the frequency of their delivery. Separate equipment should be purchased and placed in the receiving area.

Storage is either dry or refrigerated, and pages 115-117 provide a first look at dimensions for these important areas. Kitchen auxiliary areas are not always a luxury, and pages 117-118 offer suggestions for employee locker rooms and business offices. Stress the importance of these areas, although they are seldom seen by the public.

Finally, cover the very specific legal accessibility requirements: emergency exits, wheelchair ramps, railings, convenient parking, lower counters, restroom equipment, and aisles and doorways of certain widths to accommodate wheelchairs.

POINTS TO PONDER

(Reference pages for discussion questions found at the end of each chapter)

1. The simplest answer is, because it's easier to change things on paper than in a real building! Pages 99-101 mention the importance of a preliminary plan on paper – and an alternate plan, as well.

2. Pages 103-104 provide a brief discussion of patio dining, but it should be sufficient to give students an idea of its pros and cons.

3. Page 104 includes a short discussion of entryways in different types of dining environments. More details that will help students formulate their answers can be found in Chapter 2.

4. Page 105 includes a short discussion of a few factors that determine seat turnover. References to others are scattered throughout the chapter.

5. The size of a restroom depends on the seating capacity of the restaurant and, sometimes, whether alcohol is sold there. Pages 109-110 detail the requirements.

6. The dirty dish area, the dish machine, and the clean dish area; pages 112-113 explain further.

7. The receiving area is important as a checkpoint for all incoming supplies and shipments. Pages 113-115 cover the most common requirements of receiving areas.

8. Pages 115-117 discuss smart use of dry and refrigerated storage. Students can select three of these tips and expound on them.

9. This is another 'opinion question,' about the need for separate spaces for employees only. It should be interesting to see what your students have to say. Pages 117-118 offer guidelines about employee-only areas. Of course, not all restaurants have the luxury of sufficient space for such uses.

10. Page 119 discusses the need to determine a restaurant's "primary function areas." These are the spaces that must legally be made accessible according to the ADA guidelines.

QUIZ –CHAPTER 4

Circle the letter before each phrase that completes the statement correctly. If you are using a Scantron form, follow the directions on the form. Be aware that there is only one correct answer per question.

1. The method of food preparation you plan to use will help you decide:
   (a) How equipment will be used.
   (b) Which types of equipment will be necessary.
   (c) The portion sizes to be served.
   (d) The number of meals to be served.

2. Which can accurately be said about a preliminary space plan?
   (a) It requires extensive detail.
   (b) It is drawn to scale.
   (c) It does not require much detail.
   (d) It differs from the bubble diagram.

3. When planning the public areas of the restaurant:
   (a) Include some type of waiting area for customers.
   (b) Be aware that restaurants are busy no more than 20 to 25 percent of the time they are open.
   (c) Pay close attention to how you will use every booth, table, and chair.
   (d) All of the above.

4. A widely accepted standard for restaurant parking lot space is to allow:
   (a) One parking space for every 100 square feet of indoor space.
   (b) One parking space for every 50 square feet of indoor space.
   (c) Valet parking as an option, no matter what the average check.
   (d) Off-premise parking arrangements with nearby businesses.

5. If your eatery is going to offer takeout service:
   (a) Speed of service is crucial.
   (b) A varied menu is recommended.
   (c) Limit the number of pick-up stations to control foot traffic.
   (d) Add parking spaces specifically for "to go" customers.

6. One factor to always consider when planning dining room space is:
   (a) The average check amount.
   (b) The cost in dollars to construct the space.
   (c) The sizes of tables and chairs.
   (d) The daily seat turnover numbers.

7. The standard amount of space to allocate per seat in a dining area:
   (a) Is 20 square feet.
   (b) Is 10 square feet.
   (c) Is 15 square feet.
   (d) Varies according to local codes and the type of restaurant.

8. In order for a dining table to seat four people, its dimensions must be at least:
   (a) 30 inches by 30 inches.
   (b) 36 inches by 36 inches.
   (c) 30 inches by 40 inches.
   (d) Any of the above.

9. When rectangular tables are used for banquet seating, the total amount of space is divided by ____ to determine the number of tables needed.
   (a) 8
   (b) 10
   (c) 12
   (d) 15

10. One wait station (dimensions 20 inches deep by 36 inches in length) is needed for every ____ seats in a restaurant.
   (a) 50 to 75
   (b) 12 to 24
   (c) 20 to 30
   (d) 75 to 100

If a statement below is entirely true, circle the letter 'T' in front of it. If it is NOT ENTIRELY true, circle the letter 'F.' If you are using a Scantron form, follow the directions on the form.

11. T F The standard bar is 20 to 24 feet in length.

12. T F For cocktail lounge seating, allow 15 square feet per seat.

13 T F For a small (50-seat) restaurant, the restroom requires a space of just 75 feet.

14. T F Establishments that serve alcoholic beverages are required by law to have separate restrooms for men and women.

15. T F In commercial kitchens, most aisles should be 4 to 6 feet wide.

16. T F The industry norm for a commercial kitchen is that 30 percent of the space is occupied by equipment.

17. T F In the dishroom, 60 percent of the counter space is allocated for soiled dishes.

18. T F A single tank dishwasher will require 200 square feet of space.

19. T F In determining walk-in cooler space requirements, remember that one-fourth of the space will be aisles.

20. T F At least half of the entrances to a restaurant must be accessible for people with disabilities.

CHAPTER 5: ELECTRICITY AND ENERGY MANAGEMENT

TEACHING GOALS

- Impress upon students the dominant role that electrical energy plays in all foodservice settings. What functions would be impossible without it?
- Discuss the need for informing and involving employees to save energy and use it wisely. This includes understanding basic terminology; the meter-reading process; how energy is measured; billing and other common industry procedures.
- Discuss, step-by-step, the way energy is delivered to a building, and then distributed throughout the building.
- Explain how electrical systems work, and the combination of voltage options available for use. Stress the need for electrical knowledge in selecting equipment to fit the workplace.
- Familiarize students with different electric rates and costs, and how a typical bill might look. Also cover how annual utility costs can be charted.
- It's worth discussing whether alternative energy sources (wind, solar, biomass, etc.) are practical for foodservice use.
- Finally, discuss ways a foodservice business can conserve electricity. There are plenty of them!

MAJOR POINTS OF EMPHASIS

Ask students to determine how energy is used in different departments or work areas of a foodservice business. Discuss the choice between electricity and gas as energy options.

Take students through a restaurant for a walk-through energy audit, perhaps with a local utility company representative. Describe the different types of energy audits, explained on pages 128-129.

Awareness about the types and costs and energy consumption is key. Students must be able to read their own meters, decipher power bills and the different types of charges (demand charges, ratchet clauses, sales taxes, and possible exemptions, etc.) Pages 137-141 give specifics.

Basic terminology also is important. How wires work, how circuits work, and the ways power is measured (watts, amps, ohms, etc.) are covered on pages 130-134. Types of current, single-phase, and three-phase power, and voltage combinations are explained on pages 123-125.

Energy conservation efforts begin with knowing how much electricity you're using, the trying to scale back. Charting utility costs (pages 143-145) is the first step; dozens of ways to save are mentioned, starting on page 151.

Who needs an emergency power source? In many areas of the country, it could save a restaurant's business and food inventory. Guidelines and options to prompt a discussion about emergency planning are offered on pages 146-147.

The chapter ends with an extensive list of suggestions for creating a "greener" facility – some by building it from the ground up, others that can be retrofitted. It also reinforces the need to track utility use in order to chart savings progress.

POINTS TO PONDER

(Reference pages for discussion questions found at the end of each chapter)

1. Students' answers should be in question form, gleaned from the information on pages 128-129.

2. The definitions of kilowatt hour and kilovolt ampere are found on page 138. You need to know which type of electric meter you have, and meters read either kilowatt hours or kilovolt amperes.

3. These considerations are listed on page 150. They include cost, overall efficiency of the system, and what you want it to do; its size (a design consideration, since solar arrays are large); its life cycle; availability of maintenance for the system; and potential tax benefits.

4. These measurement terms are defined on page 131 – amps, volts, watts, and ohms – which is also where you'll find the simple answer to this question.

5. Explanations of why different rates are charged for electricity under different circumstances are found on page 139 (ratchet clauses) and in the section (starting on page 141) on consumption charges.

6. The energy index is a measurement of the amount of energy used to prepare and serve one meal. It is calculated by converting all forms of energy to BTUs, then dividing it by the total meal sales for the same time period. . This process is explained on page 143, and the students should find it easy to explain the benefit of arriving at this figure.

7. Fuel adjustment charges are cost fluctuations passed on to customers by a utility company. They may be a charge, or a credit, and are mentioned on page 143.

8. Page 151 begins an extensive discussion of the ways foodservice operators can save energy. It should be interesting to note which ones seem to make the biggest impression on students.

9. Yes, in case of emergency, you can shut off electricity with the Master Service Switch. You must know how to unlock it, though.

10. The primary reason for purchasing a power generator is the likelihood that you would need it. In areas where brownouts and/or severe weather are common, it's prudent to think about how to protect food in case of an outage. More details on pages 146-147.

QUIZ –CHAPTER 5

Circle the letter before each phrase that completes the statement correctly. If you are using a Scantron form, follow the directions on the form. Be aware that there is only one correct answer per question.

1. To retrofit a restaurant's electrical equipment means:
   (a) To obtain all new equipment.
   (b) To make existing equipment more efficient.
   (c) To obtain more favorable usage rates from utilities.
   (d) All of the above.

2. One way to manage a restaurant's energy is to conduct an energy audit, which many include recommendations that:
   (a) Kitchen equipment be retrofitted.
   (b) Some equipment be replaced.
   (c) Structural modifications be made to the building.
   (d) All of the above.

3. The basic unit of measurement for electrical energy consumption is the:
   (a) Volt.
   (b) Ampere.
   (c) Watt.
   (d) Ohm.

4. An electric circuit breaker "trips" or switches off when:
   (a) It receives a load higher than its capacity.
   (b) It is an open circuit.
   (c) It receives a lower load than its capacity.
   (d) None of the above.

5. On an electric bill, the ratchet clause is:
   (a) A practice by electric companies to account for demand charges.
   (b) A practice by electric companies to account for total energy charges.
   (c) A practice by the electric companies to account for fuel adjustment charges.
   (d) Based on a tariff analysis for each restaurant.

6. An electric circuit has three basic parts. They are:
   (a) The source, the wiring, the appliance.
   (b) The source, the current, and the appliance.
   (c) The source, the path, and the appliance.
   (d) The source, the insulation, and the appliance.

7. The highest voltage generated by single phase configuration is:
   (a) 125.
   (b) 480.
   (c) 240.
   (d) 250.

8. Which of these statements about alternating current (AC) is true?
   (a) It tends to overheat equipment faster than direct current (DC).
   (b) In the U.S. and Europe, AC cycles are always 60 cycles per minute.
   (c) It can be transmitted over long distances with minimal power loss.
   (d) It has a constant flow of amps and volts.

9. Heavy-duty appliances like deck ovens, with large power and load requirements
   (a) Are wired at the factory for single-phase, two-wire electricity.
   (b) Are wired at the factory for three-phase, two-wire electricity.
   (c) Are wired the same way as countertop equipment models.
   (d) None of the above.

10. If an appliance is rated at 12,000 watts and 125 volts, it will require ___ amps of electric power.
   (a) 9.6
   (b) 69
   (c) 6.9
   (d) 96

If a statement below is entirely true, circle the letter 'T' in front of it. If it is NOT ENTIRELY true, circle the letter 'F.' If you are using a Scantron form, follow the directions on the form.

11. T F *Dual voltage* equipment requires the same voltage but two different amps sources.

12. T F *Step rates* are often controlled by local or state regulatory agencies.

13 T F In many parts of the United States, electricity that is used to "process" food is exempt from sales taxes.

14. T F The *distribution panel* (also called a *main circuit panel*) directs electric power to different areas of the building.

15. T F The difference between actual charges and demand charges is known as *net metering*.

16. T F A *British Thermal Unit (BTU)* is the amount of heat required to raise the temperature of one pound of water by one degree Fahrenheit.

17. T F Five cubic feet of natural gas is equivalent to 5,000 BTU.

18. T F A 60-watt incandescent light bulb can be replaced with a 15-watt compact fluorescent bulb.

19. T F If a refrigerator's temperature reaches 50 degrees F. during a power outage, the food should be destroyed.

20. T F *Photovoltaic cells* collect wind energy and convert it to electricity.

CHAPTER 6: GAS, STEAM, AND WATER

TEACHING GOALS

- Explain the fundamentals of gas energy and common terminology. This includes the way a gas flame is created, and hazards associated with gas flames; the various types of gas burners, including the new generation of burners and particularly the infrared burner; the nature and operation of pilot lights and thermostats in gas-powered appliances.
- Describe how gas meters are read, and the application of energy-saving principles to gas-powered equipment. Discuss gas pipes and connections.
- Explain the principles of steam heat, temperature, and pressure. Introduce steam terminology, steam requirements for kitchen equipment, and common equipment problems.
- Students should learn the basics of water purity, including potential contaminants and pollutants, the differences between 'hard' and 'soft' water, and ways to improve water quality and taste.
- Emphasize the importance of water to the foodservice industry; how water is purchased and measured, including how to read water meters and water utility bills. Describe the drainage system, and potential drainage problems.
- Describe the various types of sinks found in commercial kitchen, and the need for different sinks for different purposes, particularly the importance of good hand washing stations. Discuss hot water heaters and the need for sufficiently hot water for certain uses.

MAJOR POINTS OF EMPHASIS

Gas energy in foodservice: The chart on page 164 shows how gas gets from the ground to the range top burner. Your local utility company should be able to provide a speaker who can discuss commercial gas uses, explain hot to read a gas meter and gas bill, and offer energy-saving tips for gas appliances.

On *gas ranges*, students should be able to name and describe the parts of a burner, types of burners, flame stability, and unsafe flame conditions. They should also be able to explain the uses of pilot lights, thermostats, and thermocouples. This information is covered on pages 164-169.

Emphasize the importance of installing the correct sizes of gas pipes and connections, and the need for quick-disconnect couplings in emergencies, using the information on pages 174-175.

Representatives of companies that manufacture gas and/or steam equipment are excellent potential speakers for the steam segment of this chapter, beginning on page 175. For steam equipment, students should become familiar with the basic types of steam appliances and how they work. Be sure they understand the relationship of the steam heat and temperature to steam pressure. The differences between high-pressure and low-pressure steam should be discussed.

Steam equipment has unique terminology. The roles of *converters, coils, steam traps, valves, return piping, steam boilers*, and *boiler horsepower* (BHP) should be clearly defined, along with common equipment problems and diagnoses.

Water systems are a big part of this chapter, including adequate pressure, meters, pipes, risers, fixtures, fittings, and valves. On page 192, we transition from drinking water to plumbing concerns, including fixtures for both restrooms and sinks. In these discussions, professional plumber would be a good guest speaker, as long as the information is not too technical.

Water drainage and discharge systems wrap up the water discussion, starting on page 196. Emphasize the uses of P traps and S traps, dome strainers, vents, stacks, and air gaps. Discuss the relationship of the building to its sewer system, and the legal ramifications of grease traps (installation, use, and cleaning). Mention common drainage problems, like backflow and back siphonage, introduced on page 198.

The *hot water heater* ends the chapter, starting on page 200. Familiarize students with the different water temperature needs in foodservice; how to determine the correct types and sizes of water heaters for specific needs.

POINTS TO PONDER

(Reference pages for discussion questions found at the end of each chapter)

1. The way a gas burner works is described in great detail on pages 164-169. The two components that mix to create a flame are gas and air. The air that goes into the burner can be regulated by adjusting the *air shutter*; the gas can be adjusted (on most burners, not all) by opening or closing the *orifice* through which gas flows.

2. *Flashback* is defined on page 166 as the "whoosh" or roaring sound a gas burner may make as it is lit. Reducing primary sir input to the burner is the best way to correct this condition.

3. Ways to save energy when cooking with gas can be gleaned from the maintenance tips on pages 169-170. We ask that students list three of them.

4. This answer is not spelled out neatly here in Chapter 6, but there are plenty of hints on page 176. (Steam cooking is covered in greater detail in Chapter 13.) For the moment, their impression should be that steam cooking is healthful, fast, efficient, economical, and versatile.

5. There are a number of ways to compensate for hard water, mentioned on pages 181, 183 and 186-187. Briefly, you can treat water with chemical additives; clean steam equipment frequently with a descaling agent recommended by its manufacturer; install a water-softening system; or filter water before use. In washing dishes, more detergent must also be used with hard water.

6. Since steam is pressurized, there is a natural loss of pressure when it is propelled a longer distance, and a natural increase in pressure it is propelled through a narrower pipe. So a pipe's length and diameter help determine how far and how fast the steam goes. Tables on page 180 illustrate this with actual numbers.

7. The difference between a *P trap* and an *S trap* is explained on page 196. On a sink, the trap is a curved section of pipe that retains some water, called a P trap if it drains into the wall, and an S trap is it drains into the floor.

8. Pages 193-196 detail numerous handy features of kitchen sinks. We ask students to name a describe three of them.

9. The grease trap (now the preferred term is "grease interceptor") is defined, and its function explained, on pages 197-198. It should be cleaned by a professional service company.

10. As mentioned on page 199, you use an *augur* or *snake* to unclog a floor drain, or pour clog-busting plumbing chemicals down the drain. They become clogged by debris and lack of maintenance, since they depend on gravity to work correctly.

11. Upfeed systems are described on pages 188-189.

12. A simple method for determining water heater size is explained on page 201 – estimating 5 gallons of water per guest, multiply by the maximum number of guests you would serve at a peak mealtime.

QUIZ –CHAPTER 6

Circle the letter before each phrase that completes the statement correctly. If you are using a Scantron form, follow the directions on the form. Be aware that there is only one correct answer per question.

1. The most commonly used type of gas for cooking, also known as natural gas, is:
   (a) Synthetic gas.
   (b) Methane.
   (c) Propane.
   (d) Butane.

2. In cooking, the chief benefit of natural gas is:
   (a) Its instant ability to provide intense heat.
   (b) Its lack of odor.
   (c) Its ability to easily be cleaned from pots and pans.
   (d) Its universal availability.

3. In a gas cooking appliance, the hole through which gas flows is called the:
   (a) Burner.
   (b) Pilot light.
   (c) Orifice.
   (d) Grate.

4. The most desirable color for a natural gas flame is _____, which indicates the flame is burning efficiently.
   (a) Yellow
   (b) Red
   (c) Orange
   (d) Blue

5. Ring-type gas burners are found in:
   (a) Range tops.
   (b) Range ovens.
   (c) Broilers.
   (d) Griddles.

6. This can be said about the infrared jet gas burner.
   (a) It uses more gas than other types of burners.
   (b) It uses less gas than other types of burners.
   (c) It burns a mixture of air and gas naturally.
   (d) None of the above.

7. The purpose of the thermostat in a gas-fired appliance is to:
   (a) Reduce the temperature of the burners.
   (b) Properly cook the product in the pot that rests on the burner.
   (c) Heat the appliance faster when it is turned on 'High.'
   (d) Control the supply of gas, and thus maintain the desired temperature.

8. One therm is equivalent to __________.
   (a) 100,000 BTU.
   (b) 50 cubic feet of gas.
   (c) 75 cubic feet of gas.
   (d) None of the above.

9. When water is referred to as hard, this means it contains:
   (a) Solids suspended in the water.
   (b) Higher oxygen levels.
   (c) Fluoride.
   (d) Higher quantities of minerals.

10. You can easily plug in or remove a gas-fired appliance by using:
   (a) A fixed gas pipe.
   (b) A flexible, stainless steel tube.
   (c) A thermocouple.
   (d) An orifice.

If a statement below is entirely true, circle the letter 'T' in front of it. If it is NOT ENTIRELY true, circle the letter 'F.' If you are using a Scantron form, follow the directions on the form.

11. T F *Steam* is heated water vapor.

12. T F The temperature of steam is related to its pressure – the higher the pressure, the hotter the steam.

13. T F Steam heat is the safest and most efficient way to thaw frozen foods.

14. T F A *bain marie* is a type of hot water "bath" used to keep delicate foods at safe temperatures.

15. T F Americans use about 100 gallons of water per day.

16. T F When solids are suspended in drinking water, filtration is the recommended remedy.

17. T F Limestone consists of *calcium carbonate*, which causes hardness in underground water supplies.

18. T F Surface water supplies and ground water supplies are similarly treated for potability.

19. T F Chemicals and gases are used to remove *hardness* from drinking water.

20. T F *Artesian water* is tapped from an underground source before it flows to the surface.

CHAPTER 7: DESIGN AND ENVIRONMENT

(**A note to the instructor:** this chapter contains a wide range of issues, some of which you may not typically address in your course. Lighting and noise reduction, for instance, may not be as important to you as teaching students about ventilation, air conditioning, and the kitchen exhaust system. However, the issues of employee safety and sanitation are universally accepted as important concerns. The choices of what to cover are up to you, based on the needs of your students and the focus of your course.)

TEACHING GOALS

- Describe the influence of lighting in creating a mood and/or making a space easier to use, both in front and back of house. Explain the use of color (and its relationship to lighting), and also how color can affect efficiency in a kitchen.
- Discuss both indoor and outdoor lighting. Familiarize students with how lighting is measured and other lighting-related terminology. Include energy efficiency as it relates to lighting; don't forget the use of natural light.
- Discuss the unique sound control needs of dining areas and kitchens.
- Explain the legal considerations in using recorded music in foodservice settings.
- Every aspect of heating and air conditioning should be included: terminology, how an HVAC system works, the need for ventilation, control of humidity as well as temperature, and the overall importance of air quality.
- Continue to focus on energy efficiency. Lighting and HVAC are expensive items that require maintenance and vigilance to keep them operating within a budget and not wasting energy.
- Familiarize students with the concept of commercial kitchen ventilation (CKV), including the importance and maintenance of exhaust hoods and fire protection systems.
- Indoor air quality (IAQ) issues, including grease, cigarette smoke, dishroom steam, and so on, should be part of the HVAC discussion. Be sure to check for legislative updates, as more states and municipalities ban smoking in public places.

MAJOR POINTS OF EMPHASIS

Restaurant lighting systems: Cover basic definitions (sparkle, transition zone, direct and indirect lighting, lumens, etc.), which are introduced on pages 207-209. Different types of artificial lighting, both lamps and fixtures, are described on pages 210-214. Stress the cost factors involved in lighting decisions, including different ways to control light levels.

Color: Explain its uses in setting mood, in dining areas, and kitchen work areas. A restaurant designer or interior designer may be able to elaborate. Discuss the relationship between color and lighting. Clearly delineate dining area light/color needs versus kitchen light/color needs.

Noise reduction: Students should learn that restaurant noise could work as either an advantage ('If it's busy, it must be good!') or a disadvantage ('Why go there if you can't have a conversation?') Cover basics of reverberant sound (intensity and frequency); noise control alternatives; and the unique challenges of decreasing kitchen noise. These are all described on pages 219-223.

HVAC systems: Explain individual comfort factors and the comfort zone; students should be familiar with the terminology on pages 223-226 (conditioned air, exhaust air, make-up air, etc.), as well as how the system operates. Here, an HVAC engineer or installer would be helpful, but only if you caution them not to be too technical in their presentation.

Air quality and ventilation: Cover the standards needed to maintain proper ventilation; the concept of air pressure footprints; and the debate about smoking in restaurants. You might also ask students how they'd deal with employees who want to smoke. Increasingly, however, remember that smoking is being banned in workplaces.

Commercial Kitchen Ventilation (CKV): There are specialists who design, build, and install exhaust systems. If you invite them into the classroom as speakers, caution them not to be too technical. However, students should be familiar with a balanced air system, the role of the hood or canopy, and grease extraction. A professional can also stress the importance of the built-in fire protection system. Find these details on pages 233-239.

POINTS TO PONDER

(Reference pages for discussion questions found at the end of each chapter)

1. Pages 210-215 offer numerous ways to achieve savings in lighting costs. We ask students for three of them.

2. Efficiency and efficacy both refer to how well a light fixture does its job, and both are measured in lumens per watt. *Efficiency* is the percentage of light that leaves the fixture instead of being absorbed by it. *Efficacy* is the ratio of light output to power output; it refers to how well the fixture uses electricity to convert it to light. Efficiency is the most critical, because it refers to the actual amount of light a fixture puts out.

3. Technology has improved the ballasts and the lamps themselves. Students may choose to leaf through pages 211-213 and name a few of these improvements. But the primary reason fluorescents are better now is the 1992 National Energy Policy Act (EPACT), mentioned on page 213; it required manufacturers to improve efficiency and label their products more clearly.

4. Pages 210-211 discuss LEDs and give example situations in which they might be used.

5. Ways to reduce *reverberant sound* are listed on page 221. We also mention keeping equipment and the HVAC system working properly as a noise control strategy. The use of soft convex surfaces is explained on the same page.

6. Six kitchen noise control suggestions are made on page 223. Students are asked to list three of them.

7. Pages 223-226 discuss how the HVAC system works. Our suggestion for "the first thing to be checked" is the thermostat – if it's not working properly, how can we tell if the system is working?

8. Page 233 lists numerous suggestions; we ask students for four of them. In addition, they may mention geoexchange technology (p. 227); demand control ventilation (p. 229), and even not allowing smoking.

9. A desiccant HVAC system removes moisture from the air, drying it out. This is used when outside air brought into the building contains too much moisture and needs to be dehumidified. See page 227.

10. On page 233, several adverse effects of poor kitchen ventilation are listed, in the section on

Commercial Kitchen Ventilation. Students are asked to list three.

11. A balanced system (defined on page 234) regulates the amount of fresh air introduced into a kitchen area based on a predetermined, minimum level of carbon dioxide.

12. Kitchen exhaust ducts are separate from other HVAC ducts, and connected only to the kitchen exhaust fans. This is explained on page 233.

QUIZ –CHAPTER 7

Circle the letter before each phrase that completes the statement correctly. If you are using a Scantron form, follow the directions on the form. Be aware that there is only one correct answer per question.

1. This can be said about a lighting transition zone.
   (a) It is a term used to indicate different areas of lighting.
   (b) It is a term used in describing outdoor lighting.
   (c) It is found in only in QSR settings.
   (d) All of the above.

2. Downlighting is:
   (a) A type of neon lighting.
   (b) The use of wall sconces.
   (c) Reflected lighting.
   (d) A chandelier, lighting the tables directly beneath it.

3. Power factor is a measurement of what?
   (a) The brightness in a lamp (bulb).
   (b) The efficiency of an electric device as it uses power.
   (c) The loss of brightness in a lamp (bulb).
   (d) None of the above.

4. Kitchen lighting should be selected to:
   (a) Give off the least amount of heat.
   (b) Be primarily fluorescent lamps (bulbs) for brightness.
   (c) Be able to withstand high humidity levels.
   (d) All of the above.

5. This can be said about LEDs:
   (a) They are a type of artificial lighting.
   (b) They are used in heating elements of electrical appliances.
   (c) They are also known as CFLs.
   (d) They are primarily blue in color.

6. The standard measurement for noise is:
   (a) Watts.
   (b) Frequency.
   (c) Decibels.
   (d) Lumens.

7. When selecting colors for a dining area, consider this factor:
   (a) Light tones make small spaces look larger.
   (b) The use of lighter colors helps to create a sense of intimacy.
   (c) Muted colors should be used to accent a wall or area.
   (d) Give the dominant role to more than one color on the wall.

8. The standard for brightness of kitchen lighting is:
(a) 20 foot-candles.
(b) 70 foot-candles.
(c) 50 foot-candles.
(d) None of the above.

9. As a smart noise reduction strategy in a restaurant, consider this option:
(a) Paint the acoustical ceiling tiles.
(b) Use high-pile carpeting on the floors.
(c) Install large mirrors on walls.
(d) Order tables with plate-glass tops.

10. What is the purpose of a performance rights organization?
(a) Collects fees if you play a radio station in your dining area.
(b) Works to make sure that performers and songwriters are paid for the public use of their music.
(c) Physically inspects foodservice businesses to ensure that its owners have paid the licensing fees in order to play copyrighted music.
(d) All of the above.

If a statement below is entirely true, circle the letter 'T' in front of it. If it is NOT ENTIRELY true, circle the letter 'F.' If you are using a Scantron form, follow the directions on the form.

11. T F By undercoating all kitchen work surfaces, we can reduce noise.

12. T F *Conditioned air* is air that has been filtered and released into a building.

13. T F *Opacity* is a tern used to describe the density or thickness of smoke.

14. T F *Dampers* are part of the heating system and regulate airflow.

15. T F The *refrigerant* in an air cooling system is both a liquid and a vapor.

16. T F One ton of refrigeration is equivalent to the energy required to melt one ton of ice in a 24-hour time period.

17. T F The primary *indoor air quality* issue for restaurant guests is whether smoking is allowed in a restaurant or bar.

18. T F The suggested ventilation rate for *indoor air quality* is higher for bars than it is for commercial kitchens.

19. T F A *balanced* commercial kitchen ventilation (CKV) system will remove heat, odors, grease, and moisture from the kitchen, by replacing stale air with fresh air.

20. T F A *Type II exhaust hood* is designed to collect and remove smoke, flue gases, and some of the grease produced in cooking.

CHAPTER 8: SAFETY AND SANITATION

TEACHING GOALS

- Employee safety concerns are key in this chapter, from preventing slippery floors, to fire safety instructions for workers, to proper construction of a fire protection system.
- Fire safety is key in cooking environments. Discuss requirements for automatic fire protection in kitchen exhaust systems. Explain how to avoid flash fires and emphasize the "Ten Commandments of Fire Safety."
- Discuss the importance of ergonomics in a foodservice environment, and the ergonomic considerations for designing entrances, public areas, and work areas.
- Familiarize students with flooring construction and materials, the use of floor mats, and cleaning and maintenance of floors, both in kitchen and dining areas.
- Explain and discuss sanitation rules and food handling procedures, as well as how to pinpoint food safety problems and correct them. NFSI and HACCP standards should be covered in detail.
- "Reduce, reuse, recycle" are now mainstream ideas – and even the law, in many jurisdictions. Discuss waste reduction methods and their effectiveness.

POINTS OF EMPHASIS

Fire Safety: The "Ten Commandments of Fire Safety" are listed on page 243. Invite personnel from your local fire department to discuss fire safety training tips for employees, as well as problems they may have encountered in fire inspections of foodservice facilities.

Discuss the walk-through (which starts on page 244) of a well-designed ergonomic environment. Encourage students to explore local foodservice facilities and take note of the ergonomic designs that work well.

Employee safety: Standards for work surface height and dimensions, found on page 246, should be easy for students to memorize. Emphasize the importance of selecting safe, sturdy flooring for dining areas and kitchens, and caring for it properly.

Cleaning: This chapter devotes quite a bit of space to cleaning and hand washing. Be sure students understand why this is necessary, and are clear on the difference between cleanliness and sanitation.

Food safety and sanitation: Does your school have a microbiology department? Perhaps you have a food safety expert right on campus. This discussion should include both NSFI and HACCP standards, on pages 253-257. A city health inspector who makes restaurant inspections would be another excellent contract.

In the waste management discussion, integrate these components: landfill use, source reduction, recycling, composting, and combustion. Emphasize the first step in a good waste management program is to track waste output by doing a waste audit. If your city is large enough to have a U.S. Environmental Protection Agency office, ask for a speaker on landfill use and waste reduction.

POINTS TO PONDER

(Reference pages for discussion questions found at the end of each chapter)

1. The National Fire Protection Association is the authority on this topic and sets the regulations, details of which can be found on pages 242-243.

2. Foodservice workers spend about 90 percent of their time on their feet. It is the bending over at uncomfortable angles that can cause muscle strain and back problems. Page 246 discusses how worktable heights can reduce this stress.

3. Small hand tools and equipment should be stored on two or three shelves in front of the people who will use them, no more than about 3 feet away from the standing area. Page 246 explains in greater detail.

4. Page 248 defines aggregate floors and the positive factors they provide for worker safety.

5. Table 8-2 on page 248 lists the advantages and disadvantages of different types of flooring.

6. Page 223 lists the questions to ask when selecting vacuum cleaners.

7. Pages 253-254 describe the standards of the NSFI and list many of its criteria for judging the sanitation of foodservice equipment.

8. HACCP stands for Hazard Analysis of Critical Control Points. It is a food safety system that consists of seven steps to identify and correct possible health or safety problems in a commercial kitchen. Much of the process is described on pages 254-257.

9. The process for safely washing cold foods, such as salad ingredients, is explained on page 259.

10. Once a product is cooked and pasteurized, its shelf life begins again. One-third of food-borne illnesses result from improper cooling of foods. Page 260 has details.

11. Any recycling plan should begin with a waste audit. This means keeping track of exactly what is thrown away, separating the waste into categories to determine how much can be recycled.

12. Student answers can be chosen from the many ideas on pages 261-265. Be sure they justify their conclusions about the best waste reduction tips.

QUIZ –CHAPTER 8

Circle the letter before each phrase that completes the statement correctly. If you are using a Scantron form, follow the directions on the form. Be aware that there is only one correct answer per question.

1 The most common cause of kitchen fires in restaurants is ________.
(a) Electrical malfunctions.
(b) Improper use of cleaning chemicals.
(c) Grease from the cooking process.
(d) Wood used in the cooking process.

2. Where would you find an automatic fire protection system?
(a) As an internal component of the exhaust hood.
(b) Throughout the kitchen.
(c) In the dishroom.
(d) All of the above.

3. Restaurant workers who don't pay attention to ergonomics can develop musculoskeletal disorders when performing such tasks as:
(a) Repetitive work, like chopping.
(b) Bending over a baker's bench to decorate cakes.
(c) Carrying heavy trays.
(d) All of the above.

4. A deck brush is another term for a:
(a) Chinese pig brush.
(b) Small, motorized floor scrubber.
(c) Vegetable cleaning brush.
(d) Broom.

5. When stairs are necessary in a kitchen, how tall should the *risers* be?
(a) 9 inches.
(b) 4 inches.
(c) 8 inches.
(d) 12 inches.

6. The height of a kitchen worktable should be ____ inches.
(a) 39 to 41
(b) 34 to 36
(c) 26 to 28
(d) 28 to 30

7. Aggregate is a term used to describe a substance used in manufacturing:
(a) Heavy-duty kitchen floor mats.
(b) Non-skid floor surfaces.
(c) Slip-resistant footwear.
(d) Moisture-resistant flooring.

8. Concrete is among the surfaces recommended for commercial kitchen floors:
(a) Only if it has been sealed with a water-based epoxy coating.
(b) Because it is durable and low-maintenance.
(c) Both 'a' and 'b'.
(d) Concrete is not recommended, because it's unsafe to stand on for long periods of time.

9. In purchasing commercial cleaning supplies, ________ should be avoided.
(a) Products with pH levels of 7
(b) Products that are biodegradable
(c) Products that are highly concentrated
(d) Products that contain phosphates and/or VOCs

10. Why is good hand washing a matter of good process control?
(a) Because people won't do it unless there is a process for them to follow.
(b) Because people can't be trusted to do it without a written procedure.
(c) Because it is necessary to document the process and create a baseline in order to improve on it.
(d) Because gloves are no longer considered food-safe.

If a statement below is entirely true, circle the letter 'T' in front of it. If it is NOT ENTIRELY true, circle the letter 'F.' If you are using a Scantron form, follow the directions on the form.

11. T F Entrance mats should be 12 to 15 feet long, and placed at all entrances to the restaurant.

12. T F Frozen foods are the most likely to contain food-borne pathogens.

13. T F Stainless steel is a popular material for kitchen surfaces and appliances, in part because it is rustproof.

14. T F Computerizing food and beverage inventory typically results in less waste.

15. T F Polish a stainless steel surface by wiping parallel to the *grain* of the surface.

16. T F The interiors of kitchen prep sinks are considered food contact surfaces.

17. T F Viruses are physical contaminants, which are a threat to food safety.

18. T F Line compost bins with plastic bags, even though the bags are not biodegradable.

19. T F An anti-microbial is a type of disinfectant that can prevent cross-contamination.

20. T F The most important part of hand washing for employees in a commercial kitchen is not how well they wash, but how often.

CHAPTER 9: BUYING AND INSTALLING FOODSERVICE EQUIPMENT

TEACHING GOALS

- In this chapter, students learn the search process to lead them to the right pieces of equipment for their commercial kitchen. This includes how to select, how to evaluate different offers from manufacturers (prices, useful life, warranties, tax advantages), and how to decide whether they really even *need* this piece of equipment before making a commitment to purchase it. To that end, teach them to see every equipment purchase as *an opportunity to improve profitability.* If it won't do that, do they really need it?
- Analyze the factors that determine a top-quality piece of equipment: ease of use, cleaning and maintenance, materials used to construct it, availability of accessories.
- Discuss the differences between buying new equipment, used equipment, and leasing versus owning. Students should have a thorough understanding of the Total Cost of Ownership concept.
- Emphasize the importance of writing thorough and accurate equipment specifications. Students should become familiar with what information is included in a specification sheet, and should be able to read and understand a manufacturer's spec sheet.
- Note the various certification agencies, and stress that students should look for the symbols of these groups on product literature to indicate certain standards have been met.
- Familiarize students with the types of warranties available and the merits of each, and with the trend of manufacturers' outsourcing warranty work to third-party servicing firms.
- Walk the class through a typical new appliance process: installation, start-up, safe operation, and maintenance tips.

MAJOR POINTS OF EMPHASIS

The first major point to cover is how to research an equipment purchase: what to ask and how to make the right decision. You may want to assign mock purchase decisions for different appliances to different groups of students, and ask them to do the research on costs and payment terms, depreciation, useful life, warranty options, local service availability, and so on, then present their decision to the rest of the class. Pages 271-275 provide background.

Be sure students understand that any anticipated savings are part of the costing process, as is depreciation for tax purposes.

Leasing and/or purchasing used equipment, are two important considerations. The basics are on pages 282-285. Again, use real-life examples in your town to compare the costs of new versus used restaurant equipment, and compare the costs of leasing to purchasing.

Learning to write equipment specs and understanding the roles of the certification agencies is key. Here, a classroom visit from a foodservice consultant would be very beneficial. If this is not possible, students can decipher the specs found on manufacturers' Web sites and in catalogs. They also can try writing their own specs, using the guidelines found on pages 288-291.

Ask for sample warranties from manufacturers so that students can see how technical they are, and how clearly they spell out terms. Use the warranty information on pages 295-296.

The advent of some (relatively) recent developments should make for some interesting classroom discussions topics:

- Purchasing used equipment online (Can you really outfit your restaurant on craigslist?)
- Equipment breakdown insurance (Is it necessary to spend this money if you keep up on maintenance?)
- Proprietary specifications (You mean you can't trust the designers and manufacturers not to give you what you ordered?)

POINTS TO PONDER

(Reference pages for discussion questions found at the end of each chapter)

1. Pages 271-272 list questions to ask about whether a piece of equipment is necessary to your operation, and expound on the topic.

2. The topics to cover with equipment sellers can be gleaned from the information in the section "Analyzing Equipment Purchases" (page 272) and on page 275.

3. For used equipment, pay no more than 50 percent of what it would have cost new; at auction, no more than 20 percent. These guidelines are mentioned on pages 282-284.

4. Equipment specs, and the reasons you need them, are spelled out on pages 287-290.

5. The book summarizes what equipment specs contain on pages 290-291. Students' answers to this question don't have to include that much detail, as long as they understand the concept.

6. The short answer: You must follow the terms of the warranty. Warranty precautions, such as adequate ventilation, correct power voltage, and clearly specifying warranty dates, are mentioned on pages 295-296.

7. On page 296, you'll find the list of five warranty types. We ask students to rate the importance of each of them; as an instructor, your input in that regard will be helpful to them.

8. The weight and height of an appliance, and whether it can be it on rollers or casters, determine whether an appliance should be installed on a wall, on the floor, or on a masonry base. Pages 298-299 elaborate.

9. A sample equipment information card is shown, and its function described, on page 300.

10. Nine different considerations for choosing a reliable equipment repair company are listed on pages 302-304, along with other suggestions, such as asking competitors or the local restaurant association for their recommendations. Students are asked to list three.

QUIZ –CHAPTER 9

Circle the letter before each phrase that completes the statement correctly. If you are using a Scantron form, follow the directions on the form. Be aware that there is only one correct answer per question.

1. When purchasing kitchen equipment, you're likely to pay the highest price:
   (a) When you buy from a manufacturer's catalog.
   (b) When you have it custom built.
   (c) When you buy it at auction.
   (d) When you buy at a trade show.

2. You should ask this when making a new equipment purchase.
   (a) What is the depreciation rate for the equipment?
   (b) What types of finishes are available for the appliance exterior?
   (c) Do I really need this piece of equipment now?
   (d) Any or all of the above.

3. In the lifecycle costing of new equipment, a potential purchaser should consider its:
   (a) Installation costs.
   (b) Insurance costs.
   (c) Eventual salvage value.
   (d) All of the above.

4. This can be said about the simple payback method for analyzing a new equipment purchase.
   (a) It considers annual savings.
   (b) It includes the annual depreciation rate for the type of equipment.
   (c) It takes salvage value into account.
   (d) It takes installation costs into account.

5. The Internal Revenue Service assigns a useful life of ___ years for all commercial kitchen equipment.
   (a) 5
   (b) 10
   (c) 7
   (d) 12

6. Gas-powered cooking equipment has an advantage over electric equipment because _______.
   (a) Of overall fuel efficiency.
   (b) There are no demand charges on gas bills.
   (c) Of split-rate costs.
   (d) All of the above.

7. The corrosion resistance in stainless steel is the result of its ________ content.
   (a) Chromium
   (b) Magnesium
   (c) Copper
   (d) Zinc

8. Galvanized steel is iron, coated with ________.
   (a) Nickel.
   (b) Chromium.
   (c) Zinc.
   (d) None of the above; it's not iron.

9. This certification agency has both sanitation and safety testing services.
   (a) NSFI.
   (b) UL.
   (c) FDA.
   (d) NAFEM.

10. Warranties have standard exemptions. They include:
   (a) That equipment must be correctly leveled.
   (b) The date the warranty goes into effect.
   (c) That non-factory personnel can assemble or install the equipment.
   (d) That the manufacturer is not liable for damaged goods (such as spoiled food) due to equipment malfunctions.

If a statement below is entirely true, circle the letter 'T' in front of it. If it is NOT ENTIRELY true, circle the letter 'F.' If you are using a Scantron form, follow the directions on the form.

11. T F When buying used equipment at auctions, pay no more than 20% of the price if it were new.

12. T F The bar code on equipment will ensure exact identification of what it is being ordered.

13. T F When leasing equipment, read the lease for maintenance costs that are not covered.

14. T F The *NAFEM Data Protocol (NDP)* links kitchen equipment to a central computer in the restaurant manager's office.

15. T F In the metric system, one inch equals 2.54 centimeters.

16. T F The *Specifier Identification System (SIS)* identifies, among other things, the company or person who created a particular equipment specification.

17. T F When writing the specific *classification information* for equipment 'specs,' delivery and installation instructions should be included.

18. T F In the United States, NSF International requires that foodservice equipment be certified, which means it meets a minimum set of safety and sanitation standards.

19. T F Warranties normally cover repair and/or parts replacement for one year for problems that occur as a result of faulty workmanship.

20. T F Before calling a repair technician when your oven malfunctions, be sure to first check the serial number of the oven.

CHAPTER 10: STORAGE EQUIPMENT: DRY AND REFRIGERATED

TEACHING GOALS

- Students should become familiar with receiving and storage situations in various foodservice operations. This includes an introduction to the types of equipment needed to weigh, measure, transport, and store incoming food and supplies.
- Explain how refrigeration works, including the parts of the system and the chemical refrigerants commonly used.
- Teach students how to calculate the amount of refrigerated space that will be needed for a restaurant.
- Introduce the different types of refrigerators, and cover construction tips, handy options, and accessories for each. This includes reach-in, roll-in, pass-through, and walk-in units.
- Cover the most common refrigeration units, like soft drink machines, soft-serve machines, ice makers, and dispensers, and how to maintain them.
- There is also a short section on draft beer service systems and keg storage.

MAJOR POINTS OF EMPHASIS

Emphasize to students that it is best to set up receiving and storage areas by first outlining the tasks you will need to perform there, and then deciding which types of equipment are necessary for those tasks -- scales, carts, trucks, pallets, and so on.

For storage, emphasize the ability to adjust shelving to meet changing needs, and the importance of correct height, width, and depth of shelves. Start with the information on pages 311-312. Ask manufacturers for videos on installation and use of shelf systems, particularly track shelving.

Cold storage discussion should include basic refrigeration principles: temperature reduction, heat transfer, air circulation, and humidity, found on pages 312-314, and 316-317. Both the refrigeration cycle and the refrigeration circuit must be understood. The environmental ramifications of refrigerant chemicals should also be discussed. In addition to pages 314-316, the nearest federal EPA office can provide updated rules and standards for refrigerant use.

Students must be about to discuss the basic types of commercial refrigeration units, as well as how to determine the types and sizes needed for different types of foodservice. Prompt them to pay special attention to construction details – doors, accessories, adaptability of the units – and features and options they'll be offered by manufacturers. Many of these are described on pages 317-324.

Cover the basic construction and operation of:

- carbonated beverage dispensers
- beer kegs and draft beer systems
- soft-serve machines (for ice cream and frozen drinks)
- ice makers

For ice makers, emphasize the difference between cubers and flakers, the uses for different sizes of ice cubes, and the ways to determine ice needs. Maintenance tips for all equipment should be covered. For most equipment, manufacturers' catalogs and spec sheets are readily available online.

POINTS TO PONDER

(Reference pages for discussion questions found at the end of each chapter)

1. Incoming deliveries must be checked to see whether they are correct, complete, fresh, and whether they meet the standards of the owner or chef.

2. You decide which types of scales to have in a receiving area based on what it is you need to weigh. Are there bulk items you must pay for by the pound; or big items, too heavy to be lifted onto a countertop scale? And so on. Some of it is simply a matter of personal preference. Options are described on pages 307-310.

3. Carts are rolling shelf units; trucks are platforms on wheels, designed to carry much heavier loads than carts. Both are described on page 310.

4. Temperature reduction, humidity, and air circulation combine to cool the contents of a refrigerator. These are listed on page 313 and explained further on pages 312-314.

5. Air circulation is what removes heat from foods, and from the sealed interior of a refrigerator, allowing them to cool to safe temperatures. It is accomplished with fans. The process is described on page 316.

6. Different refrigerator door options are listed on page 318 and (for reach-ins and roll-ins) on pages 319-320. They include half-doors, stainless steel or shatterproof glass, sliding doors or hinged ones, etc. Students are asked to explain their choices. One key point is to decide which way a hinged door should open, based on kitchen traffic and what else the open door may be blocking.

7. There are three methods to calculate the need for refrigerated space, found on page 322. The simplest is to allow 1 to 1.5 cubic feet of space per meal served in casual settings; 2 to 5 cubic feet per meal in fine dining settings. There are two other more complex formulas, explained in the section.

8. A dozen walk-in cooler features are listed on pages 324. Students are asked to choose three and explain why they think these are good choices. (Choices include such things as strip curtains, a foot treadle, thermometer, and heavy-duty door closer.)

9. An ice maker's capacity is rated by how many pounds of ice it can produce within 24 hours. But, as mentioned on page 328, a number of factors can alter that output. So it is safe to count on less ice than the machine manufacturer's estimate.

10. The characteristics of different types of ice cubes are listed on page 329.

11. An *inlet chiller* recirculates cold water that is already being drained from an ice machine, and uses it to begin chilling fresh water on its way into the ice machine. Its advantages are explained on page 328.

12. A *postmix* beverage machine combines syrup concentrate and carbonated water in separate, chilled tubes (lines) to create a soft drink. In a *premix* machine, the syrup and water are already mixed, and move through a single line to be dispensed. This is explained on page 325.

QUIZ –CHAPTER 10

Circle the letter before each phrase that completes the statement correctly. If you are using a Scantron form, follow the directions on the form. Be aware that there is only one correct answer per question.

1. The main reason a receiving area needs a counter scale is:
   (a) To weigh large and/or bulky items.
   (b) To account separately for the tare weight of containers and packaging.
   (c) Convenience, because they can roll.
   (d) In fact, receiving areas no longer need counter scales if they have digital scales.

2. This statement is true about electronic scales:
   (a) They should not be used to weigh items that are clearly too big for them.
   (b) They don't perform well if they get too cold.
   (c) When the readout (the scale digits) drifts, the unit may have a serious problem.
   (d) All of the above.

3. How does a cart differ from a truck?
   (a) The cart has a sturdy platform.
   (b) The truck is a rolling shelf unit.
   (c) The cart can handle heavier loads.
   (d) The truck can handle heavier loads.

4. The maximum width of a storage shelf should be ___ inches.
   (a) 12
   (b) 24
   (c) 36
   (d) 18

5. As stated in Chapter 10, a basic explanation of the principle of refrigeration is that:
   (a) It introduces cold temperatures into closed spaces.
   (b) It is the removal of heat from an enclosed space.
   (c) It is the air movement that chills food in an enclosed space.
   (d) It uses a refrigerant that remains liquid at any temperature.

6. One of the following *is not* a condition needed in refrigeration:
   (a) Temperature reduction.
   (b) Water circulation.
   (c) Air circulation.
   (d) Moisture.

7. Under which circumstance would you want a soft-serve ice cream dispenser with a high BTU rating?
   (a) When it is a gravity-feed machine.
   (b) When it is a pressurized machine.
   (c) When you want the ice cream to be thicker than normal, or you need high volume output from the machine.
   (d) When you need to maximize the amount of overrun in the mix.

8. A basic formula for determining a refrigerator's capacity is to:
   (a) Multiply the number of total square feet needed by 4.
   (b) Multiply the number of total square feet needed by 40 percent.
   (c) Multiply the number of total square feet needed by 2.5.
   (d) Estimate that only 60 percent of walk-in space will be used to store food.

9. This can be said about premix and postmix beverage dispensing systems:
   (a) The components of the beverage and the lines it goes through must be refrigerated.
   (b) The postmix syrup is more expensive than premix, because the premix is already mixed.
   (c) Neither system can be used with a bar gun or cobra gun.
   (d) All of the above.

10. An advantage of having an organized system for a foodservice receiving area results in:
   (a) Fewer pest control problems.
   (b) The ability to check incoming orders for accuracy.
   (c) A safer environment for employees.
   (d) All of the above.

If a statement below is entirely true, circle the letter 'T' in front of it. If it is NOT ENTIRELY true, circle the letter 'F.' If you are using a Scantron form, follow the directions on the form.

11. T F Relative humidity compares amounts of moisture in the air at a given temperature, expressed as a percentage.

12. T F Fresh fruits and vegetables do well in an environment of up to 80 percent humidity.

13. T F Half doors on reach-in refrigerators help to control humidity.

14. T F An ice machine should take no longer than 20 minutes to complete one ice-making cycle.

15. T F A convertible reach-in can be used as either a refrigerator or a freezer; if it has compartments with separate doors, it can be both.

16. T F An inlet chiller can boost an ice maker's capacity by 50 percent, but uses 30 percent more electricity to do so.

17. T F It's a general rule that a restaurant needs a walk-in cooler if it serves more than 250 meals a day.

18. T F An ice maker's capacity is determined by how many pounds of ice it can produce in a 24-hour period.

19. T F In a *direct draw* beer system, the beer lines must not be longer than 48 inches.

20. T F The ideal ambient temperature for an ice maker is 70 degrees Fahrenheit.

CHAPTER 11: PREPARATION EQUIPMENT: RANGES AND OVENS

TEACHING GOALS

- Discuss the basic cooking processes. Most students will have covered these in other courses, but they are important here as they relate to the ranges and ovens introduced in the chapter.
- Students should become acquainted with the range top components and details of construction, use, and maintenance of ranges. Stress the wide variety of ranges available, as well as the unique attributes of the electric and electric induction range tops.
- Students should be familiar with the equipment on this list: restaurant ranges, heavy-duty ranges, specialty ranges, range ovens, deck or stack ovens, pizza ovens, convection ovens, wood-fired ovens, mechanical ovens, rack ovens, and smoker-cookers.
- Explain the uses of proofers and retarders, impinger/conveyor ovens, rotisserie ovens, and cook-and-hold ovens.
- The microwave cooking process should be discussed, as well as its uses in foodservice.
- Discuss how purchasing decisions for ovens are made.

MAJOR POINTS OF EMPHASIS

Basic cooking processes are found on page 339. Conduction, induction, convection, and radiation should be defined.

For range tops, begin the discussion with construction details and how to determine what type of range top to order (how big, which options, which types of burners). Use the information on pages 340-347 as a starting point.

Introduce the medium-duty (also called a restaurant or café range) on page 341; and the heavy-duty range (page 342). Explain the versatility of these appliances when used in tandem with other kitchen equipment. Show photos and spec sheets for some specialty ranges, too. Manufacturers' catalogs are free for the asking.

Discussion of electric range tops (pages 345-347) should include construction details and utility requirements; the burn-in process necessary to begin using a brand new range top; and burner options, including the electric induction burner.

Discussion of the range oven begins on page 347. There are so many different oven types that we recommend inviting two or three manufacturers' reps for a panel discussion of oven technology and sales information. For each oven type listed in the textbook, students should be able to discuss its basic uses, any advantages or disadvantages, construction details, and maintenance tips.

Since microwave cookery (page 360) is so different than other types, we suggest a speaker who can speak specifically to this topic. Again, make students aware of the microwave's unique components, construction, uses, and maintenance – and focus, perhaps, on misconceptions they may have from being familiar with microwaves in home use. Wood-fired ovens and/or bakery operations are also unique enough to warrant a special presentation or field trip to see them "in action."

Finally, pages 365-366 cover the most common considerations when choosing ovens, along with new cooking technology (ask for manufacturers' demonstration CDs) and oven cleaning tips.

POINTS TO PONDER

(Reference pages for discussion questions found at the end of each chapter)

1. A *café range* (pages 341-342) is a light-duty, multipurpose range with 6 to 10 open burners. A *heavy-duty range* (page 342) looks similar, but is constructed from heavier materials and designed for long hours and high-volume cooking.

2. *Conduction* is the transfer of heat from one item to another when the two items come into direct contact. *Induction* is the transfer of heat by creating an electromagnetic field. Both are described on pages 339.

3. *French hotplates* are smaller than rectangular ones, and although both types are economical to use, the French hotplates use less power. These and other common burner types are described on page 346.

4. Page 348 lists a number of oven features. Four of them pertain to oven doors; students are asked to select three.

5. Range ovens and deck ovens work the same way, but a range oven (pages 347-348) is located beneath a range top (more like a conventional home oven), while a deck oven (page 350) stands alone or is stacked 2 or 3 ovens high.

6. Mechanical ovens are preferred in high-volume foodservice operations, especially where a single item (like bread) is produced in quantity. They are described on page 353.

7. In an impinger/conveyor oven, high-velocity hot air is forced into the oven cavity, concentrated on food that travels the length of the oven on a moving conveyor belt. The process is explained on pages 351-352.

8. Temperature control (how easy it is to get the amount of heat and smoke needed); the source of the heat (wood fired or electricity); and the size of the grill on which food is cooked are among the things to consider when selecting a smoker-cooker. This appliance is introduced on pages 355-356.

9. Home microwave ovens are not constructed for the high-volume needs of a foodservice business and, in most cities, you'd be violating a safety regulation by using a home model for commercial use. . This is explained on page 360.

10. The floor may have to be reinforced to safely hold the weight of a wood-fired oven. It must be vented correctly, and your city may also require anti-air pollution measures for the wood smoke. It's not an installation requirement, but students may mention that you should be sure there is a dependable wood supplier, too.

QUIZ –CHAPTER 11

Circle the letter before each phrase that completes the statement correctly. If you are using a Scantron form, follow the directions on the form. Be aware that there is only one correct answer per question.

1. In analyzing the power requirements for baking ovens, this factor should be considered:
   (a) The type(s) of bread to be baked.
   (b) The sizes of baking pans that will be used.
   (c) Whether the equipment is wired for single-phase usage.
   (d) Whether the equipment will be located under an exhaust hood.

2. This can be said about conduction as a cooking process.
   (a) It is the simplest form of heat transfer.
   (b) It cooks using an electromagnetic field.
   (c) It cooks using moving air inside the oven cavity.
   (d) It uses waves of energy to cook food.

3. The standard width of a commercial range is ____ inches.
   (a) 18
   (b) 36
   (c) 40
   (d) It can be any of the above.

4. One major difference between the restaurant range and the heavy-duty range is:
   (a) The type of energy used (gas or electric).
   (b) The manufacturer; some manufacturers specialize in one or the other.
   (c) The place in the kitchen where it is typically located.
   (d) The number of burners it has.

5. This can be said about the wok range.
   (a) It has adjustable rings.
   (b) Its top is shaped like a cone.
   (c) It is designed to cook large pieces of meat.
   (d) It looks like a deep fat fryer.

6. Heavy-duty electric ranges are___ inches square.
   (a) 30
   (b) 36
   (c) 40
   (d) 48

7. This can be said about an electric induction range top.
   (a) It requires an extensive heat/power source.
   (b) It has a concentric ring hot top.
   (c) It uses less energy than its electric or gas counterparts.
   (d) None of the above.

8. This can be said about a range oven.
   (a) It typically has 40,000 BTU capacity.
   (b) Its doors are hinged at the bottom.
   (c) It contains 2 inches of insulation in its walls.
   (d) All of the above.

9. This can be said about a microwave oven.
   (a) It must never be operated empty.
   (b) It heats food "from the inside out."
   (c) It is not an efficient appliance for defrosting frozen foods.
   (d) All of the above

10. The following statement is true about French hotplates:
   (a) They are associated with electric range tops.
   (b) They are used in a la carte style cooking.
   (c) They weigh less than rectangular hotplates.
   (d) All of the above.

If a statement below is entirely true, circle the letter 'T' in front of it. If it is NOT ENTIRELY true, circle the letter 'F.' If you are using a Scantron form, follow the directions on the form.

11. T F In a *conveyor oven*, the speed of the conveyor belt should be adjustable.

12. T F A pizza oven is a special application of the *deck oven*, which can reach temperatures of 700 degrees Fahrenheit.

13. T F A "cook-and-hold" oven has a maximum temperature range of 320 degrees Fahrenheit.

14. T F For long, slow-cooking uses, such as meat smoking, an electric *smoker-cooker* unit is the best choice.

15. T F Gas-fired *rotisserie* ovens also require an electric power source.

16. T F A wood/gas combination oven can be located under the exhaust hood on a hot line.

17. T F Commercial kitchens use *microwave ovens* primarily because they are small and versatile.

18. T F *Convection ovens* can accommodate any size or number of baking pans because they use conveyor belts to move food through the oven.

19. T F The appliance called a *retarder* stops yeast growth in bread dough by controlling humidity.

20. T F A *combi-wave* appliance combines microwave and deck oven capabilities.

CHAPTER 12: PREPARATION EQUIPMENT: FRYERS AND FRY STATIONS

TEACHING GOALS

- Explain the components of the deep fat fryer. Discuss operation, similarities, and differences between gas and electric fryers.
- The health implications of frying must be addressed. Students can be encouraged to research new types of cooking oil – a few are addressed in the textbook, but new alternatives are always being developed.
- Explain how fryers are installed, and discuss how to determine a fryer's capacity.
- Discuss the correct use of frying oil, and ways to conserve it. Include both manual and automatic methods of oil filtration.
- Introduce the pressure fryer, its uses, advantages, and disadvantages.
- Students should be able to name and describe each part of a fry station.
- Explain the use, construction, and care of a pasta cooker.
- Students should have basic familiarity with a few of the newer frying technologies, including greaseless fryers and computerized fryers.

MAJOR POINTS OF EMPHASIS

Begin with the basic construction of a fryer and the rudiments of the frying process. Emphasize the fryer safety tips found on page 369, and be sure students understand the major factors that impact frying oil quality: water, sediment, salt, and oxygen. Introduce concept of heat recovery time, explained on page 371.

Show students the differences between gas and infrared fryers, electric fryers and electric induction fryers. Manufacturers' representatives can come to class to describe the models they sell. Ask them to address the issues of where to locate the fryer in the kitchen, and how to choose the right capacity. There are also training materials available from equipment dealers to demonstrate fryer use and maintenance. A visit to a fast-food restaurant is recommended to see multiple fryers in action.

Students must understand the concepts of fat turnover and oil take-up, both defined on page 377. Skimmers and filters should be examined, both manual and automatic.

The pasta cooker is a specialty item, but an important one for many types of cuisine. Our discussion of it is brief (on pages 387-388) but an equipment dealer should be able to provide sales literature and spec sheets, many online. Pasta cookers are manufactured by the same companies that make fryers.

Manufacturers should also be able to contribute to the discussion on pages 385-388 about innovations in fryer technology. It is important that students see the high degree to which innovation drives the foodservice equipment industry. Ask their opinions of the new technology.

POINTS TO PONDER

(Reference pages for discussion questions found at the end of each chapter)

1. Water, sediment, salt, and oxygen – each can speed the decomposition of frying oil. These are listed on page 371. Filtering and regularly changing the oil can help prevent this, as can frying at proper temperatures, and choosing a fryer with a fast heat recovery time, and a

good-sized cold zone where crumbs and debris can drop away from the hot oil.

2. *Heat recovery time* is the time it takes for frying oil to return to its optimum cooking temperature after cold food has been dropped into it. A fast (less than 2 minute) heat recovery time prevents the food from absorbing the oil, which would make it soggy instead of frying it crisply. The term is explained on page 371.

3. The amount of food it can produce is more important than the amount of oil a fryer can hold. Capacity guidelines are discussed on page 374.

4. Fryers have a host of special installation requirements, which are explained on pages 374-376. Most are for safety, since we're dealing with hot oil, but a few are for convenience. Students might add that a fryer should be installed by an experienced installer, not a general electrician or plumber.

5. An automatic turn-down feature turns the fryer temperature down to a holding point of 190 to 275 degrees Fahrenheit when it is not being used. This helps to conserve oil, as explained on page 377.

6. A *pressure fryer* cooks food with a combination of hot oil and steam. It is fast, and the oil temperature for frying can be lower than in a conventional fryer, which saves energy. This appliance is introduced on pages 380-381.

7. A *fry station* contains everything needed to fry foods and prepare them for serving. Its common components are listed on pages 382-383. Students are asked to name four.

8. Deteriorating frying oil smells bad, foams, and can produce smoke. Oil deterioration is described on page 377.

9. A *pasta cooker* is a kettle that works like a fryer, a cabinet into which a perforated basket or kettle of food is submerged. The cabinet holds water, not frying oil, and the food to be cooked is pasta. This appliance is described on pages 387-388.

10. Capacity, simplicity (ease of use and cleaning), reliability, and size (space allocation for a fryer) are the four key points here.

QUIZ –CHAPTER 12

Circle the letter before each phrase that completes the statement correctly. If you are using a Scantron form, follow the directions on the form. Be aware that there is only one correct answer per question.

1. Trans fatty acids are a cause of heart disease. They are the result of:
   (a) The addition of chemicals to oil to keep it liquid.
   (b) The addition of hydrogen to oil to keep it liquid.
   (c) The source of the oil – animal fats instead of vegetable fats.
   (d) All of the above.

2. When frying food, oil take-up is:
   (a) High if the fryer's oil temperature is too high.
   (b) How much oil is absorbed by the food being fried.
   (c) Lower if the oil temperature is high.
   (d) Associated with the fryer's heat recovery time.

3. The size of a frying kettle is:
   (a) Stated in cubic inches.
   (b) Determined by the capacity of food it can hold.
   (c) Determined by its temperature range.
   (d) Determined by the amount of cooking oil it holds.

4. In a frying kettle, where is the crumb tray located?
   (a) On the side of the kettle.
   (b) Inside the filtration system.
   (c) At the bottom of the kettle.
   (d) Directly above the cold zone.

5. This can be said about the cold zone in a frying kettle.
   (a) It is located at the bottom of the kettle.
   (b) It keeps the kettle from overheating.
   (c) It controls the kettle's recovery time.
   (d) It holds the reserve frying oil.

6. This statement is true about fryer screens.
   (a) They are found only in electric fryers.
   (b) They are similar to, and serve the same function as, crumb trays.
   (c) They separate the cold zone from the cooking zone in a fryer.
   (d) All of the above.

7. This can be said about electric fryers.
   (a) Their heating elements are immersed in the oil inside the kettle.
   (b) They can preheat to 350 degrees F. in 6 minutes.
   (c) They require less oil than gas fryers.
   (d) All of the above.

8. The correct ratio of cooking oil to food in a fryer is ______.
   (a) 5:1.
   (b) 1:5.
   (c) 10:1.
   (d) 1:10.

9. Deep fat fryers must be located a minimum of ____ inches from other types of open-flame cooking equipment.
   (a) 12
   (b) 16
   (c) 10
   (d) There is no such requirement for separating fryers from other equipment.

10. This can be said about frying oils.
   (a) They are similar to butter.
   (b) They are lipids.
   (c) They are similar to lard.
   (d) All of the above.

If a statement below is entirely true, circle the letter 'T' in front of it. If it is NOT ENTIRELY true, circle the letter 'F.' If you are using a Scantron form, follow the directions on the form.

11. T F At room temperature, fats are in a liquid state.

12. T F The *smoke point* of frying oil is as high as 400 degrees F.

13. T F The life of frying oil can be extended by 10-15 days with proper filtration.

14. T F Frying oils last longer and produce higher quality fried foods when they are filtered every other day.

15. T F *Pressure fryers* cook with a combination of hot oil and steam.

16. T F *Food warmers* in a fry station use infrared lamps.

17. T F The chemical properties of oil change when it is heated, even without immersing food in the oil.

18. T F For greaseless frying, an electric model fryer is preferable to a gas-powered fryer.

19. T F *Ventless fryers* have internal filtration systems that capture the grease and return clean air to the kitchen.

20. T F In a *pasta cooker*, it takes one quart of water to cook one pound of product.

CHAPTER 13: PREPARATION EQUIPMENT: BROILERS, GRIDDLES, AND TILTING BRAISING PANS

TEACHING GOALS

- Present the various cooking terms associated with broilers and griddles.
- Explain the construction details, general operation, and maintenance of overhead and underfired broilers.
- Introduce students to the most common types of specialty broilers and griddles.
- Explain the construction details, general operation, and maintenance of griddles. Be specific about the proper and improper uses of the griddle's cooking surface, and how to clean it correctly.
- Describe the versatility of the tilting braising pan, including its construction details, general operation, accessories, and maintenance.

MAJOR POINTS OF EMPHASIS

Each of these pieces of equipment has several names, which may confuse students who are unfamiliar with them. Make sure they can correctly identify each basic type of broiler.

For *overhead broilers*, cover their dimensions, power requirements, and unique parts (grid assembly, grease tray). Use the cooking guide on page 395 to illustrate the differences that temperature settings and cooking times can make on food items that are commonly broiled, such as steaks.

For *charbroilers*, cover the same dimension and power information. Contrast the cooking times from page 395 to the charbroiler cooking times on page 396. Introduce the drop-in, countertop, and freestanding models.

Specialty broilers like *salamanders* and *cheesemelters* are so common that they should also be featured in classroom discussions. Sales materials, spec sheets, and catalogs are available online from manufacturers. Our discussion, on pages 399-401, also includes wood-burning broilers.

Broiler use and maintenance will be a key point of discussion. Students should know how to season and clean these appliances, and how to inspect their ventilation systems. You can add your own expertise to the basic tips found on pages 402-404.

Teach students the parts of the griddle, and explain how the concepts of heat zones and heat recovery time impact griddle-cooked foods.

Pay special attention to the griddle's cooking surface, described on pages 404-406. Discussion should include what griddles are made of, how to (and how *not* to) clean them, and griddle variations (grooved surfaces, countertop, and floor models). If manufacturers' representatives are invited to the classroom, ask them to describe as many different griddle examples as possible, their uses and advantages. We mention seven specialty products on pages 366-368.

The tilting braising pan warrants its own class period or two, just to explain its versatility. We list many of the uses for this handy appliance on pages 410-412. Students should become acquainted with typical tilting braising pan construction, dimensions, and utility requirements.

POINTS TO PONDER

(Reference pages for discussion questions found at the end of each chapter)

1. An *overfired broiler* (393-394) has its heating elements above the food and is designed for high-volume output. An *underfired broiler's* (394-398) heating elements are located beneath the food being cooked. . Its use of flames and smoke make it popular in display kitchens.

2. The *grid assembly* on a broiler is the grill on which food is placed.

3. The countertop broiler stands on a counter or a stainless steel stand. The drop-in broiler is installed in a counter. Both are small units, for restaurants that don't require high-volume food output. The freestanding broiler is a larger appliance, mounted on its own cabinet. It is usually part of a hot line or battery of cooking equipment. All of this is explained on page 398.

4. A *salamander* is a small version of broiler that can be placed above another cooking appliance or on a shelf or counter. It can be used as a broiler or plate-warmer. It is introduced on page 399.

5. The burner ports on a gas-powered broiler may be blocked; the pilot light may have been blown out by an air draft; or for electric broilers, they may not be plugged into an outlet that doesn't provide enough voltage for their full heating capacity. These are mentioned on page 402-404.

6. The griddle's heat recovery time is impacted by how cold the food is when placed on its surface. The lower the griddle's surface temperature drops, the longer it takes (and the more energy it uses) to return to optimum cooking temperature. See page 408.

7. Page 409 describes the griddle cleaning and seasoning process. Students should note that cleaning is done first, then seasoning.

8. A clamshell, introduced on page 406, is a griddle with a heated lid that cooks food on two sides simultaneously.

9. It is large enough to cook in quantity, and is a combination of the features of several different appliances, allowing it to be used in many ways. (Students may list some of these features, including a hinged cover, vertical sides, griddle surface, etc.). The full description is found on pages 410-412.

10. The type of cooking you do determines the size of tilting braising pan you will need. Different formulas are mentioned on pages 412.

QUIZ –CHAPTER 13

Circle the letter before each phrase that completes the statement correctly. If you are using a Scantron form, follow the directions on the form. Be aware that there may be no more than one correct answer per question.

1. The term griddling technically refers to:
   (a) Open-flame cooking.
   (b) Cooking on a solid top griddle.
   (c) Broiling, with heat from overhead.
   (d) Open air or outdoor cooking.

2. This is an advantage of overhead broiling:
   (a) Fast cooking times.
   (b) Slow cooking times.
   (c) Food can be broiled without smoke.
   (d) Flame broiling makes foods look attractive.

3. This statement correctly applies to the location of a drip pan:
   (a) It is above the grid assembly.
   (b) It is above the grease tray.
   (c) It is found in charbroilers.
   (d) It is found in hotel broilers.

4. This can be said about a charbroiler:
   (a) It is a type of underfired broiler.
   (b) It is often found in display kitchens.
   (c) It is an appliance that produces flames.
   (d) All of the above.

5. Radiants are found in:
   (a) Gas-fired charbroilers.
   (b) Overbroilers.
   (c) Griddles.
   (d) All of the above.

6. Reflectors are found in:
   (a) Overbroilers.
   (b) Open-hearth broilers.
   (c) Griddles.
   (d) All of the above.

7. The salamander is a type of:
   (a) Broiler.
   (b) Griddle.
   (c) Fryer.
   (d) None of the above.

8. In purchasing a broiler, check how the burners are shaped and spaced, because:
(a) This affects how even the heat will be on the cooking surface.
(b) The closer burners are together, the harder they are to clean.
(c) If they are made of lava rock, they will have to be replaced more frequently.
(d) Wavy ones retain heat better than flat ones.

9. To clean an electric broiler, you must:
(a) Soak the heating elements in the pot sink.
(b) Avoid burning off any food residue.
(c) Use oil to clean the heating elements.
(d) None of the above.

10. When radiants are part of a broiler, they may be made of ______.
(a) Copper.
(b) Nickel.
(c) Zinc.
(d) Stainless steel.

If a statement below is entirely true, circle the letter 'T' in front of it. If it is NOT ENTIRELY true, circle the letter 'F.' If you are using a Scantron form, follow the directions on the form.

11. T F The *griddle's* cooking surface is known as the *grill.*

12. T F It is not smart to place broilers and refrigerators in close proximity in a commercial kitchen.

13. T F The benefit of a *clamshell griddle* is that it can cut cooking time in half.

14. T F The *Mongolian barbecue* is another name for a *wok range.*

15. T F *Seasoning* means to lightly coat the griddle surface with sea-salt.

16. T F The cooking surface of most *tilting frying pans* is two inches thick.

17. T F A *trunnion* is used to empty liquids from a tilting frying pan.

18. T F For cooking frozen foods, a 1-inch thick *griddle plate* is recommended.

19. T F The standard depth of a *drop-in broiler* is 24 inches.

20. T F The *grid assembly* of the *overhead broiler* holds the food as it is broiled.

CHAPTER 14: STEAM COOKING EQUIPMENT

TEACHING GOALS

- Present the physical aspects of steam production and cooking, including the associated terminology.
- Describe the construction, operation, and maintenance of the steam-jacketed kettle. Illustrate the variety of steam-jacketed kettles, emphasizing how to select the right size of kettle and how to accurately determine kettle capacity.
- Explain the different ways kettles can be mounted in kitchens, and the way energy and steam are directed to the kettles. Mention the accessories that are available with most kettles.
- Describe the construction, operation, and maintenance of pressure steamers. Delineate between low-pressure, high-pressure, and pressureless steam equipment, and explain how each type is used. Include details of their dimensions, capacities, and energy requirements.
- Familiarize students with the combination steamer, the flash steamer, the needle injection steamer, the pasta and rice steamer, and the steamer/fryer.
- Spend some time describing the combination steamer/oven, its uses and versatility; and the vapor oven, along with the uses for vacuum steam.
- Explain the role of the steam generator in providing steam for cooking equipment, along with equipment cleaning and maintenance tips.

MAJOR POINTS OF EMPHASIS

Students surely know what steam is, but be certain they understand the reasons it is such an effective cooking medium, and the differences between regular steam, dry steam, and superheated steam.

Introduce the concept of a steam-jacketed kettle before you acquaint students with all the different types of kettles and their uses. There are only about a half-dozen steam equipment manufacturers, and all of them have excellent and sales and training materials to share. Take advantage of these resources.

Give students some examples of foods and amounts to be prepared, and challenge them to determine the right number and size(s) of kettle(s) to do the job. Kettle sizing guidelines are discussed on pages 420-421; manufacturers' literature should also offer this information.

Talk about where and how to install kettles in a kitchen setting. Discuss the importance of plumbing and utility availability, including steam at sufficient pressure to use in the equipment, and overall water quality issues. If necessary, refer students back to Chapter 6 for the original discussions on steam energy.

Mention the accessories available for steam equipment to maximize its versatility. We list the most common ones on pages 422-423.

Make sure students grasp the differences between regular ("pressurized") and pressureless steaming. Again, manufacturers' literature will help you explain the advantages of each type of steamer, how they are used in kitchens, their energy needs, and water quality needs. In the text, this information is found on pages 425-428.

Acquaint students with the specialty steamers listed on page 429. Trade publications regularly print articles about interesting uses of steam equipment, so be on the lookout for these.

The combi oven/steamer is described starting on page 430. Emphasize its energy efficiency and versatility. Also mention the low-temperature vacuum steam (vapor oven) option, another power-saving option.

The boiler or generator that produces steam is just as important as the cooking appliance itself. Our discussion begins on page 433.

POINTS TO PONDER

(Reference pages for discussion questions found at the end of each chapter)

1. As mentioned on page 418, the energy carried by steam is transferred immediately upon contact with a cooler surface. The steam in a steam-jacketed kettle is introduced into the space between its interior and exterior kettle walls, giving the kettle a lot more heated surface than the same size of regular pot or pan. This is explained on pages 418.

2. A steam-jacketed kettle can also be used to make coffee.

3. Yes, you can mount a floor-model kettle on a wall. It is often done for sanitation reasons, to allow cleaning beneath, as mentioned on page 422.

4. Pressure steamers require you keep the appliance's doors closed during cooking, and steam stays in the sealed area inside the appliance, not coming into contact with the food. Pressure steamers use less water and less energy than pressureless steamers. Pressureless steamers do not cook as quickly, but are easy to use. And, since the steam does come into direct contact with food, pressureless steam cooking adds moisture that improves the quality of, for instance, frozen foods.

5. Water purity is critical to produce steam for cooking. As a restaurateur, you are responsible for treating your own water to make steam. Water quality is discussed on pages 417 and 434-436.

6. Thirty two inches.

7. Whether the food is frozen or fresh is a major determinant of correct cooking time in a steamer. The amount, size, and shape of the food, as well as the depth of the food in its pan, and the type of pan used, all affect cooking time.

8. The pasta and rice steamer (page 429) would be preferred in cook-to-order situations, because it's designed to cook individual servings.

9. The high-pressure steamer is the appliance with the highest heat transfer ratio, since it holds steam under the most pressure. It is introduced on page 424.

10. This is a judgment call for students, but they are most likely to choose a pressure steamer, combination steamer, or combi oven/steamer .

QUIZ –CHAPTER 14

Circle the letter before each phrase that completes the statement correctly. If you are using a Scantron form, follow the directions on the form. Be aware that there is only one correct answer per question.

1. Which of the following appliances is best for cooking raw vegetables?
   (a) Stockpot.
   (b) Tilting frying pan.
   (c) Steamer.
   (d) Steam jacketed kettle.

2. Which of the following appliances is best for preparing soups and sauces?
   (a) Stock pot.
   (b) Tilting frying pan.
   (c) Steamer.
   (d) Steam jacketed kettle.

3. Superheated steam is water that has been heated:
   (a) To 212 degrees F.
   (b) Under pressure.
   (c) Above 212 degrees F.
   (d) None of the above.

4. To raise the temperature of steam:
   (a) Simply increase the temperature.
   (b) Simply increase the pressure.
   (c) Decrease the water level.
   (d) Use the escape valve on the steam appliance.

5. The desirable water pH factor for optimal use of steam equipment is:
   (a) 7.0.
   (b) 5.0.
   (c) 3.0.
   (d) 9.0.

6. One of the most useful features of the steam jacketed kettle is that it can:
   (a) Increase pressure rapidly.
   (b) Transfer heat rapidly.
   (c) Change temperature rapidly.
   (d) Cook food while retaining its flavor.

7. The ________ on a steam kettle allow(s) it to pivot.
   (a) Legs.
   (b) Handles.
   (c) Trunnion.
   (d) All of the above.

8. The highest possible pressure capability of a compartment steamer is ____ psi.
(a) 5
(b) 10
(c) 12
(d) 15

9. This can be said about high-pressure steamers.
(a) They are used for small batch cooking.
(b) They hold fewer pans than low-pressure steamers.
(c) They reduce cooking time compared to regular steamers.
(d) All of the above.

10. This can be said about pressureless steamers.
(a) They allow the user to increase the cooking temperature as needed.
(b) They allow the user to open the doors during cooking.
(c) They allow the user to increase the pressure during cooking.
(d) All of the above.

If a statement below is entirely true, circle the letter 'T' in front of it. If it is NOT ENTIRELY true, circle the letter 'F.' If you are using a Scantron form, follow the directions on the form.

11. T F It takes 2 BHP to run one steamer compartment.

12. T F A *vapor oven* is a good choice for cooking delicate foods, such as fish and some types of fresh vegetables.

13. T F One benefit of a *pressureless steamer* is that it is inexpensive to install.

14. T F Pressurized steam is called *lazy steam* because just "hangs" there in the steamer compartment; it is not blown around by a fan.

15. T F To heat individual portions using a steamer, consider the *flash steamer.*

16. T F The *combi-oven* is so named because it combines three cooking modes.

17. T F The *connectionless steamer* is so named because it does not need to be plugged into a power source in order to create steam.

18. T F *Combi-ovens* require that the pH factor of the water used in them be 7.0 or above.

19. T F The *vacuum steamer-cooker* uses much less energy than a conventional steamer.

20. T F Two small steamers may be preferable to one large steamer because steamers run most efficiently when they are full.

CHAPTER 15: COOK-CHILL TECHNOLOGY

TEACHING GOALS

- Define the terms *cook-chill* and *retherm.*
- Explain the cook-chill process step-by-step, identifying the equipment used to perform each step. Emphasize the importance of temperature control and accurate temperature monitoring during the process, which is the only way to ensure food safety in this process.
- Describe all food chilling options, including both blast chilling and tumble chilling processes.
- Familiarize students with the storage and distribution of cook-chilled foods.
- Focus on the ways that recipes, reheating (rethermalization), and even the ways the foods are served may all be different with cook-chill than other forms of food preparation.
- Discuss the advantages and suggested applications of cook-chill technology.

MAJOR POINTS OF EMPHASIS

Whether you call it *cook-chill* or *cook and chill*, students must have a basic understanding of this process, which has grown in popularity. The chart on page 440 is a very basic outline; the chart on page 453 goes into greater detail. Take your students to a cook-chill facility if there is one in your area. It will make every step much easier to grasp. If there is not one in your area, major manufacturers Groen (in Elk Grove Village, Illinois) and Williams Refrigeration Ltd. (in Maywood, New Jersey) have excellent training information about their cook-chill systems.

Stress the importance of small details in the cook-chill steps: selecting the best raw materials and the correct storage containers for foods, portioning foods correctly for fastest chilling. Temperature monitoring and accuracy of thermometers also are critical.

Prechill preparation is key to safe and speedy chilling. Suggestions are made on page 442. These methods, and some time constraints, are necessary to keep food at safe temperatures.

Students should know the capacities, temperature range, and space required for both blast chillers and tumble chillers; how and why each type of chiller is used. For tumble-chilled foods, acquaint students with the entire pump-fill process.

The use of conduction shelves and cook-cool tanks is also important, as well as the role of the ice builder. These options are covered on pages 446-447.

It is natural to place emphasis on the chilling process, since the equipment is unique and interesting. But do not neglect what happens AFTER chilling: how to correctly store, reheat, and distribute the food. Again, there are constant temperature requirements to maintain product safety, and rules for using product earlier if it should accidentally be thawed earlier than planned. Stress the need to monitor temperature.

Many of the advantages of cook-chill technology are listed in this chapter. Be sure students appreciate its versatility, but that they also understand its limitations: initial set-up cost, geared to high production instead of a la carte cooking, etc.

Foodservice industry publications often feature examples of successful cook-chill applications, such as the article about Beau Rivage featured in the textbook. Be on the lookout for these to share with students.

POINTS TO PONDER

(Reference pages for discussion questions found at the end of each chapter)

1. Advantages of cook-chill are mentioned on pages 439-440. Students are asked to name and describe two of these advantages.

2. Food should be stored between 32 and 38 degrees F. (page 447) – although students may point out that the illustration on page 441 lists the range for "Blast Chill Store" at 33 to 36 degrees F.

3. You should check incoming raw materials for freshness and sanitation, as with any type of food you would serve. If it hasn't been stored properly by your supplier, it won't produce the best cook-chilled product and may, in fact, be unsafe.

4. Meats should be cut into smaller pieces for cook-chilling; no more than 6 pounds, or 2-and-a-half inches in thickness. This is to facilitate faster chilling.

5. Moisture content of the food; the size, shape, and density of the food; the food's ability to retain heat; the temperature of the food; the container in which it is packed, and whether the container is covered.

6. Rethermalization means reheating, and with cook-chilled foods, it can be done in a variety of different appliances, as explained on pages 447-448.

7. How to label a cook-chilled product: type of food, preparation date, and where it is supposed to go if it has been prepared for an off-site location or one of multiple locations.

8. The cook-chilled foods may thaw too quickly in a standard refrigerator, which is opened and closed often in a commercial kitchen. You need to monitor their temperatures carefully. They may also be cross-contaminated by other, fresh foods stored with them.

9. An ice builder is an ice machine that supplies icy-cold water to a tumble chiller. Its use is discussed on pages 446-447

10. Tumble chilling (pages 445-446) is done in a rotating drum full of cold water, and is appropriate for foods that are liquid enough to be pumped into heavy-duty plastic casings and sealed. Blast chilling (pages 442-444) is done in a refrigerated chamber, best used for solid foods that sit on shelves inside the blast chiller.

QUIZ –CHAPTER 15

Circle the letter before each phrase that completes the statement correctly. If you are using a Scantron form, follow the directions on the form. Be aware that there may be no more than one correct answer per question.

1. This can be said about the cook-chill method of food preparation.
   (a) It is not the same as ice shock.
   (b) It is a system of preparing foods safely.
   (c) It prolongs shelf life of most cooked foods.
   (d) All of the above.

2. Cook-chill can prolong the useful shelf life of cooked foods by _____.
   (a) 2 days.
   (b) 5 days.
   (c) Flash-freezing them.
   (d) Cooking them to order.

3. There is general agreement that cook-chill's _________ is its biggest advantage.
   (a) Consistency
   (b) Food safety
   (c) Affordability
   (d) All of the above

4. Blast chilling is:
   (a) An appropriate method for chilling solid foods, such as beef roasts.
   (b) A more versatile quick-chilling method than tumble chilling.
   (c) Capable of reducing the temperatures of cooked foods to between 32 and 37 degrees F. within 90 minutes.
   (d) All of the above.

5. The safe storage temperature for cook-chilled foods must be reached within this time period.
   (a) 90 minutes
   (b) 30 minutes
   (c) 60 minutes
   (d) Three hours

6. When reheating cook-chilled foods, the food must be heated to this minimum temperature in order to kill bacteria.
   (a) 140° F.
   (b) 180° F.
   (c) 150° F.
   (d) 212° F.

7. When modifying recipes for use in cook-chill procedures, experts suggest the amount(s) of salt and/or sugar be ___________.
   (a) Increased by 10%.
   (b) Reduced by 25%.
   (c) Reduced by 10%.
   (d) Increased by 25%.

8. In rethermalization, it is essential that the food's core temperature reach this level for at least 2 minutes.
   (a) 150° F.
   (b) 140° F.
   (c) 180° F.
   (d) 165° F.

9. When storing cook-chilled foods, it is recommended that they be covered. Why?
   (a) Because of food safety concerns.
   (b) Covering them reduces surface dehydration.
   (c) Covering foods eliminates flavor transfer.
   (d) All of the above.

10. The difference between blast chilling and tumble chilling is that:
   (a) Blast chilling uses cold air; tumble chilling uses cold liquid.
   (b) Blast chillers have smaller motors than tumble chillers.
   (c) Blast chillers are for smaller pieces of food; tumble chillers for larger foods.
   (d) All of the above.

If a statement below is entirely true, circle the letter 'T' in front of it. If it is NOT ENTIRELY true, circle the letter 'F.' If you are using a Scantron form, follow the directions on the form.

11. T F When chilling food in standard, shallow steam table pans, the pans should not be loaded more than two inches deep.

12. T F Food should not sit for more than 45 minutes between when it is finished cooking on a rangetop and when it enters a chiller.

13. T F *Cook-freeze systems* use less energy than *cook-chill systems.*

14. T F *Tumble chillers* are usually selected to cool mostly-liquid food products, such as soups and stews, but they also can be used for some pasta dishes.

15. T F A steam jacketed kettle is a necessary piece of equipment for foods that are to be *tumble-chilled.*

16. T F An *ice builder* is a handy, but not necessary, accessory for a *tumble chiller.*

17. T F If cook-chilled food reaches a temperature of 50 degrees F., it should be discarded.

18. T F If cook-chilled food will be eaten cold, it should be consumed within 30 minutes of leaving the refrigerator.

19. T F Paper and disposable plastic plateware may be used in a *rethermalization cart.*

20. T F It is preferable to install a separate walk-in cooler for cook-chilled foods.

CHAPTER 16: DISHWASHING AND WASTE DISPOSAL

TEACHING GOALS

- Introduce the three basic components of mechanical dishwashing: hot water, detergent, and motion. Discuss the importance of each.
- Conservation has become more important than ever in dishroom operation. Make it a point to remind students about both the water and energy use implications of their decisions; also the health and environmental factors surrounding detergent use.
- Examine the search process necessary to select a dishwashing system. This includes discussion of sizes, types, layouts, and accessories for the dish machines.
- Explain the role of, and importance of, booster heaters.
- Students should be familiar with the basic construction and operation of food waste disposers, waste pulpers, and waste collectors.
- Discuss the special washing requirements for pots and pans, including manual soaking as well as mechanical washing.

MAJOR POINTS OF EMPHASIS

The essential parts of dishwashing are a sufficient amount of water at the correct temperature to sanitize dishes; good quality detergent; and adequate water circulation to remove food and stains and rinse dishes fully. Each of these factors has costs associated with it, and there are also labor costs to consider.

Be sure students understand the importance of dishroom personnel. Giving them adequate training and decent working conditions will pay off with a better-operating dishroom.

Pages 458-459 include the many variables to consider in selecting a dishwashing system. Among the first is to determine how many dishes you'll be doing; then, exactly which functions the dishroom employees have to accomplish. (Additional guidelines about the size of the dishroom itself are found in Chapter 4).

Pages 464-472 describe the half-dozen major types of dish machines. Students should be able to take this information and make dishroom decisions about their fictitious restaurants. Emphasize their need to know the machine's capacity, temperature and power requirements, ventilation requirements, and any options that are available (like blower-dryers, booster heaters, prewash features, etc.)

Students should understand the formula for determining the dish capacity of the flight-type dishwasher.

Explain the function of the booster heater, the importance of its location, and the option of using gas, electricity, or steam to power it.

Stress the restaurateur's need to pay serious attention to the waste disposal process. We began this discussion in Chapter 8, and continue here with specific equipment information about disposers, pulpers, and waste collectors. Students should know how they operate, a few desirable features, installation suggestions, and power requirements.

Finally, introduce the options for pot and pan washing, both by machine and by hand. You might ask students to see if they can calculate the cost differences for washing a certain number of pots and pans, by hand or by machine. Discuss how to select the best location for this messy job, as well as water and power requirements.

POINTS TO PONDER

(Reference pages for discussion questions found at the end of each chapter)

1. According to Table 16-2 on page 459, Joe's Cafeteria would use 10 to 12 dish items per customer (from 8,000 to 9,600) plus 800 trays.

2. We ask students to take the above figure and select a dish machine for the cafeteria. Based on the information in Table 16-4 on page 462, we must assume that the dishes for all 800 meals must be dealt with during the same time frame (the lunch rush). Joe's might be able to get by with a single-tank conveyor dishwasher, but a double- tank conveyor would be most efficient, especially considering the fact that trays must also be washed.

3. Dish machines put out a lot of heat and moisture, which can make for a very uncomfortable work environment if it is not properly vented away. This is not mentioned specifically, except to say that in some places, ventilation is required by law.

4. *Pump efficiency* is the number of gallons of water circulated through the dish machine per minute. The *rate of flow* is how fast the dish machine is refilling its storage tank. Both terms are defined on page 463.

5. The faster glasses dry, the less likely they are to emerge from the glasswasher with water spots on them. The hotter the wash water, the more quickly glasses dry.

6. A *blower-dryer* takes up a lot more room in the dish area, so you must first be able to accommodate it. It also requires venting, and greater electrical capacity. . See page 469.

7. Circular systems may take up less room than straight-line dish machines, and they require fewer people to operate them. These are described on pages 470-472.

8. The dish capacity of the machine described is 14,040 dishes per hour. The formula for calculating this is on pages 473-474.

   (120 x 5.2 = 624; 624 x 45 = 28,180; 28,180 divided by 2 = 14,040).

9. The ability of a waste disposer to reverse its motor is helpful to clear the disposer when it jams.

10. The water press in a *waste pulper* allows wastewater inside the pulper to be reused, which saves water. By squeezing water out of the slurry, it also decreases the overall waste output. Waste pulpers are introduced on page 480.

QUIZ –CHAPTER 16

Circle the letter before each phrase that completes the statement correctly. If you are using a Scantron form, follow the directions on the form. Be aware that there may be no more than one correct answer per question.

1. A detergent's wetting action refers to:
   (a) The amount of time it takes to clean dishes.
   (b) Its ability to loosen food particles.
   (c) The amount of water it requires to work properly.
   (d) The force with which the water is sprayed onto dishes.

2. The water pressure for a rinse cycle in a dish machine should be ___ psi.
   (a) 5-10
   (b) 15-25
   (c) 25-30
   (d) Any of the above; it depends on the type of machine.

3. A common and effective method for controlling the amount of dish detergent used is:
   (a) To use the right detergent.
   (b) To make dishwashers account for it at the start and end of each shift.
   (c) To pre-scrape and soak dishes before washing.
   (d) To adjust the force of the water inside the dish machine.

4. A restaurant serving about 200 guests per day should:
   (a) Use a flight-type dishwasher.
   (b) Use a single-tank, door-style dishwasher.
   (c) Use a single-tank conveyor dishwasher with a pre-wash option.
   (d) Consider its menu before selecting a dishwashing system.

5. This can be said about a booster heater:
   (a) It is used with low-temp (chemical) dishwashers.
   (b) It is used to raise water temperature from 140 to 180 degrees F. for rinsing.
   (c) It is a sanitation requirement for all commercial dishwashers.
   (d) It is an auxiliary piece of equipment that ensures the water temperature is maintained.

6. In mechanical dishwashing, there are three essential parts. Which of these *is not* considered essential?
   (a) Specific racks for different sizes of plates.
   (b) Sufficient amounts of water at proper temperatures.
   (c) The correct type and amount of detergent.
   (d) Adequate time for water recirculation.

7. The most controversial ingredient of dish detergents for its potential to pollute waterways is ___________.
   (a) Calcium.
   (b) Phosphate.
   (c) Mercury.
   (d) Magnesium.

8. Sanitizing in a mechanical dishwasher with a chlorine system:
(a) Allows the highest water temperature to be 140 degrees F.
(b) Does not require a ventilation system.
(c) Prevents the use of silverplated flatware.
(d) All of the above.

9. The minimum hot water temperature for a stationary rack, single-tank dishwasher is:
(a) 165° F.
(b) 140° F.
(c) 180° F.
(d) 195° F.

10. A benefit of low-temp dishwashing is:
(a) It uses less energy, thereby lowering costs.
(b) It allows dishes to dry more quickly.
(c) It allows a shorter overall wash cycle.
(d) The machine can be placed under the ventilation hood, saving space.

If a statement below is entirely true, circle the letter 'T' in front of it. If it is NOT ENTIRELY true, circle the letter 'F.' If you are using a Scantron form, follow the directions on the form.

11. T F The average undercounter dishwasher uses from 3 to 5 gallons of water per wash cycle.

12. T F In a *glasswasher*, the entire wash-rinse-dry cycle takes about 20 minutes.

13. T F In order to use a *blower-dryer* on a dish machine, you must have several feet of additional space and significant exhaust capacity.

14. T F The *double-tank* dishwasher can handle a maximum100 to 200 racks of dishes per hour.

15. T F The *pre-wash option* is found only in *flight-type conveyor* dishwashers, because it requires a separate motor.

16. T F The *mechanical pot/pan washer* saves on labor, water, and detergent.

17. T F *Circular dishwashing* systems use *rack-conveyor* dishwashers.

18. T F The *flight-type dishwasher* requires a special type of rack for cups and glassware to fit onto its pegged conveyor belt.

19. T F You should select a *waste disposer* based on the number of meals served each day.

20. T F A *waste pulper* can reduce the physical volume of trash but, to do so, it requires twice as much water as a *waste disposer*.

CHAPTER 17: MISCELLANEOUS KITCHEN EQUIPMENT

TEACHING GOALS

- Describe the construction, names of parts and accessories, and multiple uses of the following types of equipment:
    - Mixers
    - Food processors
    - Food slicers
    - Blenders
- Stress the safety aspects of operating each of these pieces of equipment, and provide maintenance tips to prolong their useful life.
- Introduce the different types of commercial toasters, the output and advantages of each.
- Explain the importance of keeping cooked foods at safe temperatures using food warmers and steam tables. Introduce the different types of warmers.
- Describe the major types of coffee-making machines, including espresso machines. Explain the multiple factors that are important in brewing any type of coffee.

MAJOR POINTS OF EMPHASIS

The capacities, construction, and operation details for mixers are found on pages 485-490. Also discuss important options (like benches for small mixers and bowl dollies for larger ones) that can increase the adaptability of the equipment. Show students the wide variety of mixer attachments and explain their correct uses. Discuss the concept of *absorption ratios* when mixing dough. Specialty types of mixers are introduced on pages 489.

Mixers, slicers, and processors are such integral parts of a commercial kitchen that you should bring samples of all three into the classroom, and/or ask a chef to come in and demonstrate their use.

Food slicers are described on pages 490-491; food processors on pages 491-494; and blenders on pages 494-496.

Discuss the real-life capacities of toasters, in contrast to manufacturers' estimates. Ask students to decide which toaster is best for different types of restaurants. Toasters are described on pages 496-499.

Be certain that students understand food warming equipment must meet strict legal temperature and sanitation requirements for food safety reasons. They also have specific electrical needs. Manufacturers' literature can illustrate the various types of warmers, but students should be familiar with the basic categories: overhead and base-heated, soup warmers and drawer warmers, steam tables and bain maries. Our discussion is found on pages 499-502.

In class, discuss the qualities necessary for perfectly brewed coffee. Students surely will have experience visiting – and in some cases, working in – coffee bars, so they're no doubt familiar with the potential for profitability. Ask those who do have such experience to help you introduce and demonstrate the variety of brewing appliances, including urns, airpots, and French presses. Include details about power needs and plumbing requirements. Our "coffee break" begins on page 502.

POINTS TO PONDER

(Reference pages for discussion questions found at the end of each chapter)

1. The motor on a commercial mixer is located on or near the upper housing of he mixer. This is among the mixer "construction facts" mentioned on page 487.

2. You use a *bowl adapter* when you're putting a smaller bowl on a mixer than its usual size of bowl. It holds the bowl firmly in place, as described on page 486.

3. A *slicer* can't slice if its gauge plate is set at 'zero,' because that means making slices with 'zero' thickness. On most slicers, you cannot detach the carriage from the slicer unless the gauge plate is set at zero, for safety reasons. This is explained on page 490.

4. Cheese is easier to grate in a food processor if it comes right out of the refrigerator and is dusted lightly with cornstarch to prevent sticking. This is mentioned on page 492.

5. A blender performs light-duty mixing jobs. It contains a blade assembly that whirs at the bottom of the blender container, which holds the food (usually a liquid) being blended. A spindle blender's mixing head is on a metal arm, longer than the blade assembly of a typical blender, is placed in the container from the top instead of the bottom. Both blender types are described on pages 494-496.

6. Vertical conveyor toasters are much larger than horizontal conveyor toasters. You select one based on your amount of available space, since their output is similar. Page 499 introduces these toasters.

7. As seen on page 499, the minimum holding temperature for food is 135 degrees F., so the food warmer should be able to hold food at this internal temperature.

8. Keeping pans of food covered as they sit in a steam table conserves heat and energy.

9. Coffee brewers (pages 503-504) make coffee by filtering hot water through a basket of ground coffee beans. An *espresso machine* (page 505) is a specialty appliance that brews small batches of strong coffee and makes high-pressure steam to foam milk to add to the coffee. It may also grind coffee beans.

QUIZ –CHAPTER 17

Circle the letter before each phrase that completes the statement correctly. If you are using a Scantron form, follow the directions on the form. Be aware that there is only one correct answer per question.

1. A food mixer is identified by the __________.
   (a) Size of its motor.
   (b) Speed of the beaters.
   (c) Size of the mixing bowl.
   (d) Number of mixing speeds available.

2. The standard 20-quart mixer found in foodservice kitchens is power by a __ horsepower (HP) motor.
   (a) One-half
   (b) One
   (c) Two
   (d) Three

3. Which of these statements is true about a food warmer?
   (a) It must provide a heating temperature of 140 degrees Fahrenheit.
   (b) It uses special warmer lamps.
   (c) It may have thermal shelves.
   (d) All of the above.

4. An electric food slicer is identified by the ____________.
   (a) Size of its motor.
   (b) Diameter of its blade or knife.
   (c) Speed of its blade.
   (d) Number of slices that are produced per minute.

5. Which of the following statements is true about the bowl-style food processor?
   (a) It has a variable speed switch.
   (b) It should always have a pulse function.
   (c) It has a 20-quart capacity bowl.
   (d) It can be hand washed carefully, by immersing it in water.

6. This can be said about a bar mixer:
   (a) It is similar to a blender.
   (b) It can crush ice cubes.
   (c) It has a one-pint capacity.
   (d) It is similar to a spindle blender.

7. An example of an agitator is the:
   (a) Dough arm.
   (b) Bowl dolly.
   (c) Chopping blade.
   (d) Dicer.

8. One of the following does not belong in the group with the other toasters.
   (a) Pop-down toaster.
   (b) Manual lift toaster.
   (c) Batch toaster.
   (d) Drawer toaster.

9. A combination-style food processor with a one-half HP motor can process ______ of food in one hour.
   (a) 50 pounds
   (b) 100 pounds
   (c) 250 pounds
   (d) 350 pounds

10. Cleaning is critical for maintenance of a food slicer. This includes:
   (a) Lubricating the blade with food-grade oil.
   (b) Not using bleach on the aluminum parts of the slicer.
   (c) Using clean, cut-resistant gloves.
   (d) All of the above.

If a statement below is entirely true, circle the letter 'T' in front of it. If it is NOT ENTIRELY true, circle the letter 'F.' If you are using a Scantron form, follow the directions on the form.

11. T F Day-old bread makes better toast, and it is also smart use of the product.

12. T F The higher the *absorption ratio* of dough, the harder the motor of a mixer must work to mix it.

13. T F The *continuous toaster* is capable of toasting 150 slices of bread per hour.

14. T F In commercial kitchens, the use of *bowl guards* is a federal safety requirement in the United States.

15. T F The minimum holding temperature for *food warmers* is 165 degrees F.

16. T F To keep food warm for more than one hour, a *drawer warmer* is recommended.

17. T F A *hot food table* heats hot air; a *steam table* heats hot water.

18. T F Water filtration is necessary for espresso machines, both for flavor and machine maintenance reasons.

19. T F One pound of ground coffee beans can brew from 1.75 to 3 gallons of coffee, depending on the desired strength of the brew.

20. T F *Spindle blenders* have transmissions in their motors that require oil.

CHAPTER 18: SMALLWARE FOR KITCHENS

TEACHING GOALS

- Emphasize the commonsense reasons behind the use of the correct tool for the job: portion control, recipe accuracy, and ease of the task itself for the worker who performs it. Sometimes, using the right tool is even a safety issue.
- Show students the various hand tools used in every commercial kitchen. They should become familiar with:
    - Serving spoons
    - Spatulas
    - Hand whips
    - Tongs
    - Food turners
    - Kitchen forks
    - Ladles and scoops
- Kitchen knives should be a major part of the discussion, as students must be familiar with the many types and sizes and their common names, plus how to care for them.
- Food safety and sanitation is critical here. This includes the correct use and sanitizing of cutting boards, and extensive discussion of how to take food temperatures and what types of thermometers and/or thermocouples to use for different foods.
- Miscellaneous (but important!) kitchen gadgets should be shown and discussed. . The list includes:
    - Kitchen scales
    - Strainers and colanders
    - Volume measures
- Common types of cookware, their names, shapes, uses, and construction details should be examined, as well as ovenware and bakeware items, and serving/holding containers.

MAJOR POINTS OF EMPHASIS

Hand tools are often designed for specific kitchen tasks, and using the wrong one can be unsafe, or result in too much waste or uneven portions. Point out the relationship of something simple, like ladle size, and its importance to control costs in a restaurant; or the use of color-coded cutting boards to prevent cross-contamination of foods.

Discussion of the most common types of hand tools begins on page 508. Just using one example, like the serving spoon in all its permutations, should show students how versatile these tools can be. Show students the differences in strainers and colanders, including the specialty items like China caps and food mills.

There are several good training CDs available about knife construction, use, and care. Most of them are for sale, not manufacturers' giveaways. Students must be familiar with the most common types of kitchen knives, be able to name their parts, discuss how they are made, sharpening options, and cutting board use and sanitation.

Measuring tools include kitchen scales, volume measures, and thermometers. Stress the importance of keeping scales and thermometers accurate, and introduce any technological advancements in measurement. The thermocouple and infrared thermometers are good examples.

Get price information and ask students to decide if the cost is worth the results when choosing a more technologically advanced piece of measuring equipment.

The North American Association of Food Equipment Manufacturers (NAFEM) has excellent guidelines for starter packages of smallware, created for different types of foodservice operations. These can be used with NAFEM permission.

Pot and pan manufacturers have excellent brochures about their product lines to share. Go over them in class as a group. Students should know the basic differences between a pot and a pan; the variety of metals used in cookware construction (advantages and disadvantages); and the names and uses of the following items:

Saucepot	Saucepan
Stewpot	Sauté pan
Double Broiler	Satoir
Brazier	Sauteuse

Similarly, ovenware (also called bakeware) includes a wide variety of items for different uses. Again, rely on manufacturers' literature and/or samples to show roasting pans, roast-and-bake pans, and sheet pans; and specialty items like pizza pans and springform pans.

The requirements of display and storage are different than cooking and baking. Containers must look decent, and also protect the stored foods. A cafeteria and/or catering operation would be a great place to see the variety and versatility of steam table pans and serving containers.

POINTS TO PONDER

(Reference pages for discussion questions found at the end of each chapter)

1. Spoons are used for stirring, but *spoodles* can be used for both stirring and portioning specific amounts of food.

2. A complete description of knife construction and quality features is found on pages 513-515. A top quality knife would have, among other things, a *taper ground* blade made of high-carbon stainless steel; a *tang* that runs the entire length of the handle; and a *bolster* that is an extension of the blade, not a separate metal piece.

3. Cladding is defined on page 524; it's important because the metals used affect the heat conductivity of the pot or pan, and its ability to heat evenly.

4. The list of five important considerations for kitchen scale purchase is listed on page 517. Students are asked to choose three of them, and to explain why they are important.

5. Ladles and scoops are both used to portion foods, but ladles are used primarily to transfer liquids between containers, and scoops are used mostly to pick up dry ingredients or ice. Both are described on page 518. (Again, students need to *see* these items so they can fully understand their uses.)

6. A foodservice business should have stem-type thermometers (for checking internal food temperatures); dial thermometers (for checking internal temperatures of appliances); thermometers to check dishwasher water temperature and fryer oil temperature; a cooling thermometer (in cook-chilling, this gauges the temperature drop during the quick-chill

process). Those are just a few! The full description is found on pages 519-522. (You may give students a number, so they'll know how many types to mention, or written answers may be long and involved.)

7. A nonstick coating on pots and pans does prevent some foods from sticking, and means you can use less oil in some forms of cooking. But the coatings may scratch off easily if the pan comes into contact with metal utensils or abrasive scouring pads. Details on pages 524-525.

8. A satoir is a straight-sided, shallow sauté pan. A sauteuse is a shallow skillet with gently sloping sides. They are both described on page 526.

9. Different foods come out best on different types of sheet pans, and half-sized pans are nice for displaying foods. The various sizes and uses are explained on pages 528.

10. The seasoning process is described on page 531. It involves cleaning and drying the pan, heating it until it begins to smoke, then adding salt to the pan. Roast the salt, then empty the pan.

QUIZ –CHAPTER 18

Circle the letter before each phrase that completes the statement correctly. If you are using a Scantron form, follow the directions on the form. Be aware that there is only one correct answer per question.

1. This can be said about serving spoons.
   (a) They are made of several different materials, including plastic and stainless steel.
   (b) The spoon's bowl can be solid, slotted, or perforated.
   (c) The spoon's handle is usually made of wood.
   (d) Most have a capacity of two ounces.

2. All of the following are types of spatulas, except the _________.
   (a) Spoodle.
   (b) Hamburger turner.
   (c) Pie server.
   (d Docker.

3. You would use a piano whip when:
   (a) You are whipping a liquid that is light or thin in consistency.
   (b) You need a longer whip.
   (c) You need a whip with flatter coils to get into the edges of a pot or pan.
   (d) The food you are stirring is heavy in consistency.

4. Which of the following *is not* a type of *strainer*?
   (a) The sieve.
   (b) The portioner.
   (c) The chinois.
   (d) The colander.

5. Which of the following *is not* part of a *cook's knife*?
   (a) The tang.
   (b) The spine.
   (c) The shank.
   (d) The stamp.

6. The best way to minimize food safety hazards associated with cutting boards is to:
   (a) Have plenty of cutting boards available, and color-code them for use with different types of foods.
   (b) Sanitize them in a microwave oven between uses.
   (c) Use cutting boards made of hard maple.
   (d) All of the above.

7. In cookware construction, these are the two basic factors that govern its ability to cook evenly:
   (a) The temperature retention and flavor transfer properties of the metal.
   (b) The thickness and flavor retention properties of the metal.
   (c) The type of metal used and its thickness.
   (d) The type of metal used and its temperature distribution.

8. If you were catering at a different location than your kitchen, you'd need ______ to monitor cold food temperatures in the coolers or ice chests.
   (a) Time-temperature indicators
   (b) Thermistors
   (c) Bimetallic thermometers
   (d) Environmental thermometers

9. When a knife is sharpened, what's really happening is that:
   (a) It is being ground down slightly by the sharpening steel.
   (b) The blade is being aligned by the sharpening steel.
   (c) The sharpening steel is shaving off rust and discoloration.
   (d) The taper of the bolster is being altered slightly.

10. A baker's scale can be adjusted to measure amounts in:
   (a) 1 ounce.
   (b) 0.5 ounce.
   (c) 0.25 ounce.
   (d) Ounces or centimeters.

If a statement below is entirely true, circle the letter 'T' in front of it. If it is NOT ENTIRELY true, circle the letter 'F.' If you are using a Scantron form, follow the directions on the form.

11. T F The difference between a *saucepot* and a *saucepan* is that they have different types of handles.

12. T F The capacity of a *ladle* is, in most cases, stamped on its handle.

13. T F *Spatula/scrapers* should be made of *thermoplastic* so they won't melt.

14. T F A *kitchen fork* has only two tines.

15. T F A knife with a *hollow-ground blade* is preferable for foodservice use.

16. T F A full-sized *sheet pan* measures 18 by 26 inches.

17. T F Water boils at slightly different temperatures at different altitudes.

18. T F A *springform pan* is used to bake cheesecakes.

19. T F The *steam table pan* (also called a *hotel pan*) has a standard size of 12 by 20 inches.

20. T F Pots and pans can be washed in a commercial dishwasher, as long as the proper type of detergent is used.

CHAPTER 19: TABLEWARE

TEACHING GOALS

- Emphasize the dual function of tableware: functionality (usefulness as utensils), and marketing (reinforcing ambience and concept).
- Define a few basic terms, including place setting, tabletop, table setting, plateware, flatware, and glassware.
- Discuss tabletop trends and fashions, and stress the importance of choosing tabletop items appropriate to concept, budget, and menu. Industry publications often focus on trends in tabletop design – keep them handy to use in class.
- Familiarize students with the following:
    - The plateware manufacturing process; basic shapes; care and cleaning tips.
    - The glassware manufacturing process; basic shapes, care and cleaning tips.
    - The flatware manufacturing process; basic sizes, and how to purchase, clean, and polish flatware.
- Cover manufacturers' guidelines for how much plateware, glassware, and flatware to purchase.

MAJOR POINTS OF EMPHASIS

Ask students to dine out at different places and report to the class on tabletop presentation. They can research the prices of the items and/or interview restaurant owners about the reasons for their choices.

Students should become familiar with the basics of plate construction and firing, shapes, and sizes.

Stress the importance of correct ways to carry, bus, and store dishes to prevent chipping and breakage, in both dining room and dishroom. Emphasize the rule for glass storage: from rack, to table, to rack!

Talk about cost considerations: not just prices of items, but their overall durability and availability, and the difference between custom-made and open-stock designs. These topics are discussed throughout the chapter.

Students should know basic glassware terminology and how processes like *annealing* and *tempering* affect durability.

There are so many shapes and sizes of glassware (including wine glasses) that the sheer variety deserves special attention. . Again, our discussion is just a beginning.

Flatware has a colorful and interesting history, which should not be overlooked. Familiarize the class with basic manufacturing processes and finishes. Be sure to cover the quality tips, as well as care and cleaning guidelines. Also spend some time on the many types of forks, spoons, and knives – and help students decide what is really necessary for various types of dining experiences. The dining scene has become so much more informal – is it a good thing that few people know (or care about) how to use an oyster fork or a fish knife? What do your students think?

POINTS TO PONDER

(Reference pages for discussion questions found at the end of each chapter)

1. Tableware enhances the concept or personality of your restaurant; changing tableware is an expensive way to redecorate; and it makes foods and beverages look their best, increasing their appeal to diners. More on pages 537-539.

2. *Vitrification* is a super-heating process that makes china stronger. It is described on page 541.

3. In a fancy restaurant, the table would probably be set with a dinner plate (an oversized charger is optional); salad plate or bowl; bread plate; at least two forks (salad, dinner, optional oyster or dessert fork); three spoons (teaspoon, soup spoon, after-dinner spoon); dinner knife; butter knife; water glass; and at least one wine goblet. This answer is not spelled out in a single page or chart, but students can use the flatware chart on page 555 for guidelines.

4. Plates can be stacked, but no more than 12 inches high, on nonmagnetic, stainless steel shelves, as mentioned on page 543. Store plates at or below the food plating area, since more accidents occur when people try to lift plates from overhead.

5. Tempered glass undergoes one additional quick temperature change during manufacturing, increasing its strength. When the whole glass is tempered, it's called *fully tempered*; when only the rim is tempered, it's *rim tempered*. .

6. To prevent thermal shock, don't put ice into glasses still warm from the dishwasher, and before pouring a hot drink into a glass, preheat it under warm water. These tips, and others, are found on page 549.

7. Stainless steel flatware is popular for most foodservice uses because it is strong and durable. Silverplate is more expensive, and appropriate for upscale dining, because it is coated (plated) with silver to mimic the look of pure silver utensils. Flatware manufacturing processes are introduced on pages 551-553.

8. Guidelines for presoaking flatware are found on page 553. Limit presoaking to 30 minutes, in a low concentration of mild dish detergent. Presoaking should begin as soon as possible after utensils have been used.

9. The American exchange is the tendency of Americans to put their dining fork in one hand to hold food as they cut it on their plates, then to transfer the fork to the other hand to eat it. Oversized flatware makes this slightly more difficult, as mentioned on page 555.

10. This question requires students' personal opinions about whether to use a tabletop designer.

QUIZ –CHAPTER 19

Circle the letter before each phrase that completes the statement correctly. If you are using a Scantron form, follow the directions on the form. Be aware that there is only one correct answer per question.

1. For a six-ounce wine pour, the current trend is to use a ___-ounce wine glass.
   (a) 10
   (b) 8
   (c) 12
   (d) Any of the above.

2. In selecting tableware, you should always consider its ___________.
   (a) Durability.
   (b) Availability of replacement stock.
   (c) Appearance.
   (d) All of the above.

3. The most important point on any dining table:
   (a) Is the exact center of the table.
   (b) Is the 24-inch space directly in front of the guest.
   (c) Depends on the size and/or style of the table.
   (d) None of the above.

4. According to the National Restaurant Association, the useful life of a plate:
   (a) Is 1 year.
   (b) Is 2 years.
   (c) Is 5 years.
   (d) Depends on the quality of the plate.

5. Only one of the following *does not* affect what type of dinner plate you choose:
   (a) Whether the dishes will be exposed to extreme heat and/or cold.
   (b) The formality of the place setting.
   (c) The height of the table.
   (d) The type of flatware you use.

6. The advantage of two-fired plates is that:
   (a) They can be stacked atop each other without scratching.
   (b) Since they are fired twice, they are stronger.
   (c) They have the look and feel of china, but are more sturdy.
   (d) Their colors last longer and stand up to the rigors of commercial dishwashing.

7. It is important that dirty dishes:
   (a) Are prerinsed in the hottest possible water.
   (b) Are never stored in dishwasher racks.
   (c) Are washed within 30 to 40 minutes after their first rinse.
   (d) All of the above.

8. Which of these statements is true about the process of tempering glasses?
(a) It includes cooling the surface of the glass.
(b) It is a way to build the resistance of the glass to temperature extremes.
(c) It can be done on both stemware and tumblers.
(d) All of the above.

9. To prevent spotting when glasses are being washed, experts suggest:
(a) Towel-drying them.
(b) A wetting agent in the detergent.
(c) Drying them with higher heat than plateware.
(d) All of the above.

10. This can be said about presoaking flatware:
(a) It should not be done for more than 30 minutes.
(b) It is not necessary if the dishwasher is working properly.
(c) It can tarnish or corrode the flatware if not done carefully.
(d) All of the above.

If a statement below is entirely true, circle the letter 'T' in front of it. If it is NOT ENTIRELY true, circle the letter 'F.' If you are using a Scantron form, follow the directions on the form.

11. T F A *footed glass* is another name for a *mug*.

12. T F Wine glasses, no matter what their size, should always be clear.

13. T F Air drying is not recommended for beer glasses.

14. T F The single most important attribute of any type of tableware is that it is clean.

15. T F Soiled plates should be placed upside down in a *bus box*.

16. T F *Pyramiding* is preferable to storing glassware in dishracks.

17. T F *Burnishers* and *cutlery polishers* are small, specialty dishwashers made for cleaning flatware.

18. T F The weight of a piece of flatware is not necessarily an indication of its quality or strength.

19. T F The accepted standard for commercial flatware is that it contains 74 percent stainless steel.

20. T F The current trend is to limit glassware to a few basic shapes and sizes, to save money and enable better inventory control.

CHAPTER 20: LINENS AND TABLE COVERINGS

TEACHING GOALS

- Discuss the role linens play in the dining room. Include color combinations and common types of tablecloths: runners, overlays, deco-mats, etc.
- Explain how fabrics are manufactured. Discuss the best choices for table linens and the advantages/disadvantages of each of the following:
    - Cotton
    - Nylon
    - Linen
    - Acetate
    - Rayon
    - Blends
    - Satin
    - Damask
    - Momie cloth
- Elaborate on the major considerations in selecting table coverings. Some of these relate to concept, menu, and atmosphere; others to the qualities of the specific fabrics or materials.
- Mention industry standards for sizes of tablecloths and napkins. Look at interesting options other than linens for tabletops: paper; laminated designs or inlaid memorabilia; glass.
- Present napery care and cleaning as a step-by-step process, including its correct use (not to wipe up spills or double as potholders); correct procedures to wash, dry, iron, fold, and store.
- Especially important: troubleshooting common laundry-related problems, and most common laundry drying problems; and formulas for how many tablecloths and napkins to order.
- Discuss the costs, benefits, and potential hassles of these three options: hiring a laundry service, setting up an on-premise laundry, and linen rental.

MAJOR POINTS OF EMPHASIS

Whatever covers your tabletop, it should have both marketing and customer comfort functions. Customer comfort enters into the discussion of qualities of individual fabrics. To that end, walk students through linen decisions that include each of these considerations: fashion, budget, environment, tactile sensations, lifestyle (of your customers), menu, and technology. Other major points: whether or not you'll have on-premise laundry facilities, and the qualities of the fabric itself. Other practical factors to influence fabric choices are listed on pages 559-563, such as color retention and absorbency. The interesting thing about this discussion is that no one fabric is a "standout, must-use" – each has its advantages and disadvantages that must be weighed.

Students should become familiar with how to order linens, and in which sizes for which purposes.

Ask students to research (in your market) the cost differences, advantages, and disadvantages of linen rental; buying linen and hiring a laundry service; and installing an on-site laundry room. Cover the topics of dimensions, utility requirements, and special construction needs (non-skid floors, waterproof surfaces, shelves, and folding tables, etc.).

POINTS TO PONDER

(Reference pages for discussion questions found at the end of each chapter)

1. You'd use a *top* (also called an *overlay*) in addition to a regular tablecloth to make bussing tables easier and less noisy, and/or to add an accent color to table linens.

2. Blended fabrics can be stronger, more colorfast, and shrink proof. However, they can also absorb oily stains, and are just as susceptible to heat, bleach, and mildew discolorations as other fabrics.

3. The costs of linen rental, from energy surcharges to replacement charges to flat fee monthly bills, are noted on pages 566-567.

4. A tablecloth selection guide is printed on page 565. You choose the correct size by measuring the table and adding 8 to 12 inches for the *drop* over the sides, as explained on pages 564-565.

5. The rule of three is a linen quantity guide, described on page 565. It says for every piece of napery on a table, you should have two more in-house (one in the laundry and one in inventory). Students are also asked for their opinions of the rule; some may suggest interesting alternatives.

6. Students may complain that there's not enough information in this question to correctly prepare an answer. True, we did not mention the number of chairs at each table, nor the number of turns for each meal. Both of these factors would significantly impact the numbers of linens needed. With this in mind, ask students to assume that there are four chairs per table, and (with two meals per day) there are only two turns of each table. If this is the case, the place requires 510 tablecloths (85 x 2 x 3) and 2,040 napkins (85 x 4 x 2 x 3). We apologize for any raucous debates this question may cause!

7. Paper products, laminated tabletops, tiles, glass, and vinyl are several non-linen tabletop alternatives. They're discussed on pages 568-569.

8. After the initial investment, you can save up to 50 percent of what it costs to rent linens by buying your own and washing them yourself. You must think of it as a long-term benefit, however. With your own laundry room, you can purchase less expensive napery and hire less skilled personnel to do the work.

9. An extractor is a machine that spins laundry almost dry, allowing it to dry faster in a commercial dryer, or to go directly from the extractor to an ironer to be pressed while still damp. Sometimes, it's part of the commercial washer.

10. A list of ten common drying mistakes is found on page 577. Students are asked to mention any three of these points.

QUIZ –CHAPTER 20

Circle the letter before each phrase that completes the statement correctly. If you are using a Scantron form, follow the directions on the form. Be aware that there is only one correct answer per question.

1. The term napery in foodservice applies to:
   (a) Tablecloths.
   (b) Tabletops or overlays.
   (c) Napkins.
   (d) All of the above.

2. When guests are surveyed about what they expect from napkins, their top priority has been:
   (a) Environmental concerns – cloth can be reused, paper cannot.
   (b) The way the fabric feels – and they prefer natural fibers.
   (c) That they have enough of them for a polite dining experience.
   (d) All of the above.

3. The 50/50 blend is a synthetic textile made of these two fabrics:
   (a) Cotton and polyester.
   (b) Linen and polyester.
   (c) Nylon and polyester.
   (d) Rayon and nylon.

4. When using a linen rental service, you are likely to encounter this type of charge on the bill:
   (a) Replacement charge.
   (b) Energy surcharge.
   (c) A flat, weekly fee based on the amount being rented.
   (d) All of the above.

5. When purchasing tablecloths for a dining room, the rule of three is based on the number of tables and __________.
   (a) Seats in the dining room.
   (b) Total covers served.
   (c) The daily turnover rate.
   (d) All of the above.

6. The standard 36-inch square table requires:
   (a) A 52-inch square tablecloth.
   (b) A 60-inch square tablecloth.
   (c) A 43-inch square tablecloth.
   (d) An 11-square inch drop.

7. When designing an on-premise laundry room, don't overlook details such as:
   (a) Sufficient shelving and storage space.
   (b) The ceiling height.
   (c) The use of vapor-proof light fixtures and exhaust fans.
   (d) All of the above.

8. You can help control the disappearance of rented linens by:
   (a) Issuing bar towels and aprons by shift, on a daily basis.
   (b) Urging wait staff to not use cloth napkins to wipe down tables.
   (c) Knotting and setting aside the damaged linens for repair.
   (d) Charging the staff for replacement costs when linens go missing.

9. It is important to check the alkalinity of your restaurant's water for laundry use because this will help you determine:
   (a) How much bleach to use.
   (b) Whether you'll need to use a water softener.
   (c) The overall life of the napery fabric.
   (d) How much dirt or lint is redeposited on the fabric during a wash cycle.

10. The term hand, as it applies to fabric, refers to:
   (a) Its usefulness.
   (b) Its style or fashion.
   (c) The way it feels to a person who touches it.
   (d) Its environmental "handprint."

If a statement below is entirely true, circle the letter 'T' in front of it. If it is NOT ENTIRELY true, circle the letter 'F.' If you are using a Scantron form, follow the directions on the form.

11. T F A restaurant that uses cloth napkins should have a total of 9 to 12 of them for every table in its dining area.

12. T F When renting linens, an employee must be designated to count incoming shipments and note discrepancies.

13. T F A napkin will become a *rag-out* after about 100 washes.

14. T F *Polyester* is not biodegradable, which may be an environmental concern if your restaurant is touting a "green" image.

15. T F Most cloth napkins and tablecloths can be cut to order, to your specifications.

16. T F *Vinyl* tablecloths are thin and, therefore, generally not sturdy enough for use in outdoor dining.

17. T F The suggested laundry temperature for washing table linens is between 140 and 160 degrees F.

18. T F *Soap curd* is usually the result of a detergent that contains too much alkaline.

19. T F Experts suggest that laundry costs can be reduced by 50 percent if a foodservice business purchases and launders its own linens.

20. T F The correct size of a tablecloth's *drop* is between 8 and 12 inches.

ACTION PLAN
FACILITIES DESIGN and EQUIPMENT/FURNITURE/FIXTURES

PURPOSE: The Action Plan is a way for each student to apply the ideas and concepts in the course to the design of a specific facility of the student's choice. In order to make comparisons, it's best to limit the choices to four types of facilities, which still leaves students plenty of room for creativity:

1. A 140-seat table-service restaurant
2. A café/bakery with 40-50 seats
3. A 250-seat, single-line cafeteria
4. A QSR, 1400 square feet, with both indoor seating and drive-through window

The entire project is based on 100% of whatever grade the instructor allocates for this learning activity, with a percentage of the total grade assigned to each activity.

ACTIVITIES: Each student is expected to cover the following areas:

(a) History of the cuisine or region (5% of total grade)
(b) Concept development (10% of total grade)
(c) Proposed menu (10% of total grade)
(d) Foodservice facilities program (15% of total grade)
(e) Layout of the facility (30% of total grade)
(f) Equipment specifications and costs (30% of the grade)

SEQUENCE OF ACTIVITIES:

(A) Write a one to two page narrative about the cuisine of the proposed facility. Its development up to the present time as well as how well this type of cuisine is represented in your market. Also mention as to why you think this type of cuisine will succeed in your market.

(B) Write a two page narrative explaining your specific concept:
- Develop the concept statement
- Describe how this facility relates to common foodservice concepts
- Identify its target markets
- List the days and hours of operation
- Describe and discuss how food will be served
- Based on your space, determine how many employees you will need
- If you decide to include a bar or lounge, discuss the implications and how it will look, including choices of chairs, tables and bar
- List and explain the permits that will be required from your municipal or state government, including their cost
- For all facilities in this Action plan (except the QSR) desserts and breads will be made on-premise. Be sure to address this issue in this activity. (Review Chapter 1 to guide you in the development of this activity.)

(Use attached FACILITY DESIGN WORKSHEET)

(C) Develop a menu that includes at least 10 entrées.
- Discuss intended price levels

- Justify prices by discussing cost of foods used

(D) Develop a Foodservice Facilities Program
- Do this for every area in the facility (see attached)
- Create a "bubble diagram," as discussed in Chapter 3
- Use the guidelines in Chapters 2 through 7 to guide you through this activity, and relate your finished work to the points made in those chapters
(Use FOODSERVICE FACILITIES PROGRAM WORKSHEET)

(E) Using a 24- x 36-inch graph paper, create a layout of the proposed facility.
- Use the scale of ¼" = 1".
- Begin by deciding the broad outline of areas such as entrance(s), dining room(s), kitchen(s), bar, restrooms, storage areas.
- To fine-tune the diagram, use tracing paper to block out the major areas to size. Transfer it to the graph paper only when you are satisfied with it on tracing paper.
- Be sure to show all doors, windows, aisles, and enclosed spaces.
- Show locations of all drains and the grease interceptor (trap). Show electrical outlets and direct electrical connections, particularly in the kitchen.

Some students may choose to complete step "E" on computer. The authors have had particular success with a software program called "SmartDraw" (www.SMARTDRAW.com).

(F) Based on the schematic developed in (E), show the location of each piece of equipment.
- Number each piece of equipment on the schematic to create an equipment key. (There are two sample designs in the book, between Chapters __-__ and __-__, which include schematics and equipment keys.)
- Write and attach specifications for each piece of equipment (See Chapter 9, and use the EQUIPMENT SPECIFICATIONS WORKSHEET in this manual.)

On the NAFEM Web site, gather sample specifications as follows:
- Click on "Manufacturers"
- Click on "NAFEMIQ"
- Click on "Find Manufacturers," and select a category of equipment to browse
- Then click on "View Web site" (of the selected Manufacturer)
- On the individual sites, you'll generally click on "Products," and can access the specifications for specific pieces of equipment.

(G) From the information gathered through the NAFEM Web site develop an equipment budget.
- Be sure to list all pieces of equipment, furniture, and fixtures by area.
- For the purposes of this project, the price as listed is adequate. Do not be concerned with the myriad "additions" – such items as discounts, freight charges, insurance, warranty costs, and so on.

ACTION PLAN - FACILITY DESIGN WORKSHEET

Type of Facility: ____

1. Space

(a) Size of building lot______________ (b) Square footage of building_______________

Space Allocation

(c) Dining room(s)____________% (d) Private dining areas______________%

(e) Kitchen__________________% (f) Bar or lounge___________________%

(g) Restrooms________________% (h) Office_________________________%

(i) General storage____________% (j) Common area(s)_________________%

(k) Bar storage _______________% (l) Other__________________________%

2. Projected Sales

Food_____________% Alcoholic beverages__________%

Occupancy

	Total number served	**Seat turns**	**Required seats**
Breakfast	_______________	___________	____________
Lunch	_______________	_________	_____________
Dinner	_______________	__________	_____________

3. **Service Style(s)**

Table service______________ Counter service___________

Banquet service____________ Cafeteria ________________

Take-out service___________ Scrambled cafeteria__________

Food court________________ Curbside service_______________

Explain any particulars not covered in the general guidelines shown above: _________________

____

____

____

ACTION PLAN - FOODSERVICE DESIGN WORKSHEET
(Complete a different sheet for each area of the facility)

Name of area or space:____

Square footage required:____________________________

Description of Area: (Here you describe the use of the space, including HVAC requirements and utility requirements. Also the floor, walls and ceiling treatments, as well as the lighting requirements.)

Relationship to other spaces: (Which other areas are adjacent to this space? Describe the access to it from other spaces. The bubble diagram should help you decide these relationships.)

Special design requirements: (Here you discuss unusual or unique space requirements, including accessibility for disabled patrons or staff members, as well as structural issues.)

ACTION PLAN - EQUIPMENT SPECIFICATIONS WORKSHEET

1. Your requirements

(a) Which menu products will be *prepared/held/stored* using this piece of equipment?

(1)____

(2)____

(3)____

(4) ____

(b) How will food be prepared?

Batch cook________ Cooked-to-order_______

If cooking to order, how many individual orders will be needed at peak volume?

(1)________________________(2)________________________

(3)________________________(4)________________________

(c) Available power? Gas: Natural_____ Propane_____

Electricity:

______120V 1PH	________AMPS
______208V 1PH	________AMPS
______208V 3PH	________AMPS
______ 220V 1PH	________AMPS
______ 220V 3PH	________AMPS

Other: ______Volts __PH ________AMPS

(d) Available space: ___________Width _________ Depth _________Height

(e) Which other equipment will be located near/adjacent to this piece of equipment?

(a)__________________________(b)__________________________

(c)__________________________(d)__________________________

(f) Are there doorways to go through to get this piece of equipment from the receiving area to its final location?

________(H) X _______ (W) No doorways __________
________(H) X ________(W)

2. Availability of Equipment (This information is gathered from the specification sheets that you obtain from dealers and/or manufacturers' Web sites.)

(a) Power

Gas: ______BTU

______Natural

______Propane

Connection: _______Front

_______Rear

Electric: _________Watts

________Volts

________Phase

________Amps

Connection: ______Cord and plug

_______Hard-wired

(b) Size

(1) Overall size:

___________Depth

___________Width

___________ Height

(2) Weight:

Shipping weight_________

Uncrated weight_________

(c) Configuration

(1) Floor Model______ (2) Counter Model_______ (3) Legs_______ (4) Casters_______

(d) Finish/Housing

(1) ______Stainless Steel (2) ______Galvanized metal (3) ______Aluminum

(4)_______Other (Describe) ____

(e) Options and accessories (Here you list particular options and/or accessories you want, based on information gathered from the specification sheets.)

____
____
____
____

SUGGESTED RESOURCES AND READING LISTS

Journal/Magazine Titles	Web Sites
Chain Leader	www.chainleader.com
Cheers	www.mondotimes.com
Chef	www.chefmagazine.com
Club Magazine	www.club-mgmt.com
Consulting and Specifying Engineer	www.csemag.com
Cooking For Profit	www.cookingforprofit.com
Cornell Hospitality Quarterly	www.cqx.sagepub.com
Equipment Solutions	www.equipsolu.com
Food Service Equipment Reports	www.fermag.com
Food Arts	www.foodarts.com
Food Management	www.food-management.com
Food Service Equipment and Supplies	www.fesmag.com
Hospitality Technology	www.htmagazine.com
Hotels	www.hotelsmag.com
Journal of Forensic Science	www.journal.astm.org/jofs
Living	www.livingmagazine.net
Lodging	www.lodgingmagazine.com
NAFEM in Print	www.nafem.org
Nightclubs and Bars	www.nightclub.com

Restaurant Business	www.restaurantbiz.com
Restaurant Hospitality	www.restaurant-hospitality.com
Restaurant Startup and Growth	www.RestaurantOwner.com
Restaurants and Institutions	www.rimag.com
Santé	www.santemagazine.com
The Consultant	www.fcsi.org
The New York Times	www.nytimes.com

Author(s)	Book Title	Publisher	Year
	Restaurant and Bar '911'	Vantage Point Books, Inc.	2007
AH&LA	Energy Management and Conservation Guide	American Hotel and Lodging Association	1999
Almanza, Barbara, et al	Foodservice Planning: Layout, Design, and Equipment, 4th Ed.	Prentice Hall	2007
Asensio, Paco	Ultimate Restaurant Design	Neuse Publishing Company	2000
Baraban, Regina, S. et al	Successful Restaurant Design, 2nd Ed.	John Wiley and Sons	2007
Belasco, Warren	Meals to Come: A History of the Future of Food	University of California Press	2007
Belknap, Marie	The Restaurant and Bar English/Spanish Dictionary	John Wiley and Sons	2001
Berry, Wendell	The Unsettling of America - Culture and Agriculture	Sierra Club Books	1995
Birchfield, John	Design and Layout of Foodservice Facilities, 3rd Ed.	John Wiley and Sons	2006
Borsenik, Frank, D. et al	Management of Maintenance and Engineering Systems in the Hospitality Industry, 4th Ed.	John Wiley and Sons	2000
Broto, Charles	Club Design	Links International	2007
Casado, Matt	Conversational Spanish for Hospitality Managers and Supervisors	John Wiley and Sons	2001
Cheny, Kelly	Industrial Interiors: Bars and Restaurants	RotoVision (England)	2005
Culinary Institute of America	The Professional Chef's Knife Kit	John Wiley and Sons	2006
Del Valle, Cristina	New Bars and Restaurants	Collins Design	2005
DiPilla, Steve	Slip and Falls Prevention - A Practical Handbook	Lewis Publishing	2003
Doody, Sue, et al	As the Tables Turn	Orange Frazier Press	2006
Farrelly, Lorraine	Bar and Restaurant Interior Structures	John Wiley and Sons	2004
FCSI-EAME	Professional Kitchen	Food Service Consultants Society International	2007
Fullen, Sharron, L.	Restaurant Design: Designing, Construction, and Renovating A Foodservice	Atlantic Publishing	2002
Gdula, Steven	The Warmest Room in the House	Bloomberg	2007
Hasanovic, A., et al	100 of the World's Best Bars	Image Publishing Corp.	2006

Hendee, Susan, S, et al	Handheld Computers for Chefs	John Wiley and Sons	2007
Horn, Sam	Pop! Stand Out in Any Crowd	Penguin Trade Group	2006
Jones, Terri	Culinary Calculations, 2nd Ed.	John Wiley and Sons	2007
Katz, S., E.	The Revolution will Not Be Microwaved	Chelsea Green Publishing	2006
Kunz, Martin, N	Café and Restaurant Design	Neues Publishing Company	2006
Le Cordon Bleu	Kitchen Essentials	John Wiley and Sons	2000
Lenciani, Patrick	Death by Meeting	Jossey-Bass	2005
Loving, Amory	Winning the Oil Endgame	Rocky Mountain Institute	2005
Lynn, Jaquelyn	Start Your Own Restaurant and Five Other Businesses	Entrepreneur Press	2006
Marvin, Bill	How to Run a Business	Hospitality Management Press	2001
Meyer, Danny	Setting the Table: The Transforming Power of Hospitality in Business	Harper Collins	2006
Miller. Daniel	Starting A Small Business	Harvard Common Press	2006
NAFEM	An Introduction to the Foodservice Industry	North America Food Equipment Manufacturers	2005
NRA	A Guide to Preparing a Restaurant Business Plan	National Restaurant Association	1996
Pegler, Martin	Contemporary Restaurants and Bars	Visual Reference Publications	2004
Pegler, Martin	Designing the World's Best Bars	Visual Reference Publications	2005
Pollan, Michael	The Omnivore's Dilemma: A History of Four Meals	Penguin Press	2006
Ryder, B.	Restaurant Design	Abbeville Press	2004
Simon, Lee	The Restaurant Dream	Atlantic Publishing	2006
Stevens, J. and Scriven, C.	Food Equipment Facts	C.H.I.P.S	
Stevens, J., and Scriven, C.	Manual of Equipment and Design, 2nd Ed.	C.H.I.P.S	1999
Stipanuk, David M., et al	Hospitality Facilities Management and Design, 3rd Ed.	Educational Institute-AH&LA	2006
Sweeney, Kep	The New Restaurant Entrepreneur	Dearborn Trade Publishing	2004

Wansink, Brian	Mindless Eating: Why We Eat More than We Think	Bantam Books	2007
Wiggers, Eugene, G.	Catering to Every Whim	Prentice Hall	1991
Yabuka, N. et al	Hip Interiors: Bars and Restaurants	Visual Reference Publications	2005
Yee, Roger	Hotel and Restaurant Design	Visual Reference Publications Group	2005

ANSWER KEY

CHAPTER 1 QUIZ

1.	A	11.	T
2.	D	12.	T
3.	C	13.	F
4.	C	14.	T
5.	B	15.	T
6.	D	16.	T
7.	C	17.	T
8.	A	18.	F
9.	B	19.	T
10.	A	20.	T

CHAPTER 2 QUIZ

1.	B	11.	F
2.	A	12.	T
3.	C	13.	T
4.	A	14.	T
5.	D	15.	F
6.	C	16.	F
7.	D	17.	F
8.	A	18.	T
9.	B	19.	T
10.	B	20.	T

CHAPTER 3 QUIZ

1.	C	11.	F
2.	A	12.	T
3.	C	13.	T
4.	A	14.	T
5.	B	15.	T
6.	A	16.	T
7.	D	17.	F
8.	C	18.	T
9.	B	19.	T
10.	D	20.	T

CHAPTER 4 QUIZ

1.	B	11.	T
2.	C	12.	F
3.	D	13.	F
4.	A	14.	T
5.	A	15.	T
6.	D	16.	T
7.	C	17.	T
8.	B	18.	F
9.	A	19.	F
10.	C	20.	T

CHAPTER 5 QUIZ

1.	B	11.	F
2.	D	12.	T
3.	C	13.	T
4.	A	14.	T
5.	A	15.	F
6.	C	16.	T
7.	D	17.	T
8.	C	18.	T
9.	B	19.	T
10.	B	20.	F

CHAPTER 6 QUIZ

1.	B	11.	T
2.	A	12.	T
3.	C	13.	T
4.	D	14.	T
5.	A	15.	T
6.	B	16.	T
7.	D	17.	T
8.	A	18.	F
9.	D	19.	F
10.	B	20.	T

CHAPTER 7 QUIZ

1.	A	11.	T
2.	D	12.	F
3.	B	13.	T
4.	D	14.	T
5.	A	15.	T
6.	C	16.	T
7.	A	17.	T
8.	C	18.	T
9.	B	19.	T
10.	D	20.	F

CHAPTER 8 QUIZ

1.	C	11.	T
2.	A	12.	F
3.	D	13.	F
4.	D	14.	T
5.	C	15.	T
6.	A	16.	T
7.	C	17.	F
8.	C	18.	T
9.	D	19.	F
10.	C	20.	T

CHAPTER 9 QUIZ

1.	B	11.	T
2.	C	12.	T
3.	D	13.	T
4.	A	14.	T
5.	C	15.	T
6.	B	16.	T
7.	A	17.	F
8.	C	18.	F
9.	D	19.	T
10.	D	20.	F

CHAPTER 10 QUIZ

1.	A	11.	T
2.	D	12.	F
3.	D	13.	F
4.	B	14.	T
5.	B	15.	T
6.	B	16.	F
7.	C	17.	T
8.	D	18.	T
9.	A	19.	F
10.	D	20.	T

CHAPTER 11 QUIZ

1.	C	11.	T
2.	A	12.	T
3.	B	13.	F
4.	D	14.	T
5.	A	15.	T
6.	B	16.	F
7.	C	17.	T
8.	D	18.	F
9.	A	19.	T
10.	D	20.	F

CHAPTER 12 QUIZ

1.	B	11.	F
2.	B	12.	T
3.	D	13.	T
4.	C	14.	F
5.	A	15.	T
6.	C	16.	T
7.	D	17.	T
8.	A	18.	T
9.	B	19.	T
10.	B	20.	F

CHAPTER 13 QUIZ

1.	B	11.	F
2.	C	12.	T
3.	D	13.	T
4.	D	14.	F
5.	A	15.	F
6.	B	16.	F
7.	A	17.	T
8.	A	18.	T
9.	D	19.	T
10.	D	20.	T

CHAPTER 14 QUIZ

1.	C	11.	F
2.	D	12.	T
3.	C	13.	T
4.	B	14.	F
5.	A	15.	F
6.	B	16.	T
7.	C	17.	F
8.	D	18.	T
9.	D	19.	T
10.	B	20.	T

CHAPTER 15 QUIZ

1.	C	11.	T
2.	B	12.	F
3.	A	13.	T
4.	D	14.	T
5.	A	15.	T
6.	C	16.	F
7.	B	17.	T
8.	D	18.	T
9.	D	19.	F
10.	A	20.	T

CHAPTER 16 QUIZ

1.	B	11.	T
2.	B	12.	T
3.	C	13.	T
4.	C	14.	F
5.	D	15.	F
6.	A	16.	T
7.	B	17.	T
8.	D	18.	T
9.	A	19.	T
10.	A	20.	F

CHAPTER 17 QUIZ

1.	C	11.	T
2.	A	12.	F
3.	D	13.	T
4.	B	14.	T
5.	B	15.	F
6.	B	16.	T
7.	A	17.	T
8.	A	18.	T
9.	D	19.	T
10.	D	20.	T

CHAPTER 18 QUIZ

1.	B	11.	T
2.	A	12.	T
3.	A	13.	F
4.	B	14.	T
5.	D	15.	F
6.	C	16.	T
7.	C	17.	T
8.	A	18.	T
9.	B	19.	T
10.	C	20.	F

CHAPTER 19 QUIZ

1.	A	11.	F
2.	D	12.	T
3.	B	13.	F
4.	A	14.	T
5.	C	15.	T
6.	A	16.	F
7.	C	17.	F
8.	D	18.	T
9.	B	19.	T
10.	A	20.	T

CHAPTER 20 QUIZ

1.	D	11.	F
2.	B	12.	T
3.	A	13.	T
4.	D	14.	T
5.	C	15.	T
6.	A	16.	F
7.	D	17.	T
8.	A	18.	F
9.	B	19.	T
10.	C	20.	T

Made in United States
North Haven, CT
10 November 2026

22096550R00063